AF777795

GROUNDIE

The Indispensable and Diverse Role of the Arboricultural Ground Worker

JEFF JEPSON

Illustrations by Bryan Kotwica

Beaver Tree PUBLISHING

Groundie

Copyright © 2021 by Jeff Jepson
Published by Beaver Tree Publishing
Email: beavertree@arvig.net
Web: BeaverTreeMN.com

Third printing, April 2023

This book incorporates content originally included in *To Fell a Tree* (© 2009) and *Knots at Work* (© 2013) by Jeff Jepson and published by Beaver Tree Publishing.

Excerpt(s) from *In the Kingdom of Ice: The Grand and Terrible Polar Voyage of the USS Jeanette* by Hampton Sides, copyright © 2014 by Hampton Sides. Used by permission of Doubleday, an imprint of the Knopf Doubleday Publishing Group, a division of Penguin Random House LLC. All rights reserved.

Excerpt from *Norwegian Wood* by Lars Mytting. Text copyright © 2011 by Kagge Forlag, AG. English translation copyright by Robert Ferguson. Used by permission of Abrams Image, an imprint of Harry N. Abrams, Inc., New York. All rights reserved.

Excerpt from *Master and Commander* quoted with the permission of the authors.

All rights reserved. No part of this work may be reproduced without written permission from the publisher, except by a reviewer who may quote brief passages or reproduce illustrations in a review with appropriate credits; nor may any part of this book be reproduced, stored in a retrieval system, or transmitted in any form or by any means—electronic, mechanical, photocopying, recording, or other—without written permission from the publisher, except as provided by USA copyright law.

Disclaimer
Serious injury or death could result from the use of techniques and equipment described in this book. It is the reader's responsibility to seek qualified instruction on the techniques of ground operations in tree work. Every person involved in production tree work should use good judgment and common sense while practicing new techniques. Unfamiliar techniques should be practiced in a controlled environment before they are incorporated into everyday work procedures. This book is sold with no liability to the author, editor, publisher, or critiques, expressed or implied, in the case of injury or death to the purchaser or reader.

Cover design: Andrea Baumann
Cover illustration: Richard Solomon Artists, Douglas Smith
Interior illustrations: Bryan Kotwica.

Printed in the United States of America
Library of Congress Control Number: 2020925368
ISBN-13: 978-0-9726679-2-0

It is not the critic who counts; not the man who points out how the strong man stumbles, or where the doer of deeds could have done them better. The credit belongs to the man who is actually in the arena, whose face is marred by dust and sweat and blood; who strives valiantly...who spends himself in a worthy cause; who at the best knows in the end the triumph of high achievement, and who at the worst, if he fails, at least fails while daring greatly, so that his place shall never be with those cold and timid souls who neither know victory nor defeat.

—THEODORE ROOSEVELT

Acknowledgements

"Ideas are a dime a dozen," so the saying goes. That's probably true, because like opinions, everyone has one. However, some ideas are worth more than others. Those are the ones that get pursued and accomplished. Such was the case for this book when friend and fellow arborist Charley Wagner had the idea for a book devoted specifically for the ground worker. Four years later the idea finally became "flesh" in the form of the bound paper and ink that you're now holding. But not without the help, experience, insights, and encouragement from a host of others.

I'd like to thank each one of these people for their contribution to the book, who in some way made it better than it would have been without them: Duane Brown, Peter Gerstenberger, Sharon Lilly, Peter Jenkins, Jerry Beranek, Jay Maier, Chip and Jean Borkenhagen, and Tenlee Lund; my crew of groundies and climbers—past and present: Luke Jepson, John Klein, Koldon Oliver, Nate Linville, Bernie Fashingbauer, Ben Jordan, and Peter Held; and my daughter Anna and son-in-law Serge for their encouragement and prayer.

I especially want to thank the gals that run the office for Beaver Tree Service, without whom the business would be complete confusion and I'd have no time for writing books: Shelly Fanth, and my wife Bonnie. Thank you!

Thank you to Bryan Kotwica whose work as a professional tree worker makes him the perfect illustrator for a book such as this; Douglas Smith for the stunning cover illustration; and Andrea Baumann for once again designing a cover that captures the attention and imagination of anyone who has eyes to see.

Most of all, I thank God; the One who gives us work in the first place, the ability to perform it, and the capacity to enjoy it.

Contents

Introduction

In the tree care profession it seems that climbers get all the glory. This should come as no surprise really. After all, they're the ones dangling from the tops of trees running one of the most dangerous tools known to humankind—the chain saw—within inches of the very cordage that secures them aloft. This, and the fact that most of their daring feats are highly visible to the spectators below, it makes perfect sense that climbers receive the amount of praise and attention they do.

While the role of the groundie may not *appear* as glamorous as that of the climber, don't be fooled—it's just as daring, just as dangerous, just as challenging, and just as important. Which is to say, the ground worker is indispensable to the safety and success of production tree work. So much so, that if ground workers were ever to go on strike, most tree climbers would probably quit and most tree companies would go out of business. Even one of my proof readers observed their worth when he said, "They don't get paid enough to do what they do."

Unfortunately, the ground worker's contribution to tree work is not always appreciated to the degree that it should be. So in the same way that *The Tree Climber's Companion* was a tribute of sorts to the worker aloft, this book is a tribute to the men and women who work on the ground. More practically though, this book is a training manual—even part employee handbook and travel guide—that will take you on a journey through a day in the life of the arboricultural ground worker. From the shop to the job site, you will learn the skills necessary for becoming a great groundie. Just remember, no matter where you want to end up in the tree care industry, it all begins working as a groundie.

That said, this book is not just for groundies. It's for anyone who grabs a rope to lower a limb, picks up a saw to cut a log, or seizes a rake to clean a lawn—climber, crew leader, or boss. No matter how many years someone devotes to being a tree worker, they never "arrive" at knowing it all—learning is a lifelong journey. There is something in this book for everyone.

Before you begin reading this book, let me offer a few helpful suggestions and one final comment.

First. I've been told by many people (and to a certain extent personally observed) that young people today don't like to read much and that they probably wouldn't read this book. For those of you of whom this might be true, let me challenge you: prove me, us, them wrong! Read this book! It won't take as long as you think. Once you do finish it, start reading it all over again. You'll learn something new each time you do. This is good advice—advice I learned *from a book* published in 1904 titled, *Traditional Blacksmithing,* by J.G. Homstrom:

> "Let it be a practice or a habit, if you will, to buy at least one book every year, and to read the same, once, twice, thrice, or until its contents are indelibly impressed upon your mind."

May this be the one book that you read over and over "until its contents are indelibly impressed upon your mind." Doing so will help you make faster progress in becoming a great groundie. Besides that, you just might discover the joy of reading and go on to read some of the recommended books listed in the appendix on page 173.

Second. The little numbers that appear occasionally at the end of a sentence are there to direct you to the *End Notes* on page 176. These notes offer extra information about something that was said in the sentence.

Third. Use the indexes found inside the front and back covers to locate important sidebar articles and work skill topics. You can even record and track your learning progress by marking the adjacent check box when you've finished reading the article or section. Refer also to the *appendices* in the back pages for more information: more knots, glossary, recommended reading, and the story of one woman's experience in the tree care profession.

Final comment. In many ways this is a strange book. Lots of quotes and sidebar articles (some *seemingly* unrelated to tree work), lessons on morality and even military references. The only way I can explain it is that as a lover of books myself, I simply wrote the kind of book that I enjoy reading—one I would have enjoyed when I first started as a tree worker. Strange or not, I hope that you will enjoy the book also.

1

The Role of the Ground Worker...

At a Glance

This could be the most difficult job you'll ever have. But, if you stick with it long enough, it might become the best job you'll ever have. Sadly, most ground workers don't last very long. The majority quit within the first couple years of employment. You may be inclined to do likewise, even after only a few weeks of work. This is perfectly understandable, for tree work is hard work. It is physically, mentally, and emotional demanding. It is also extremely dangerous work. The tree care profession consistently ranks near the top of the list of most dangerous occupations. It's no wonder then why so many move on to find safer and softer work.

But for the rugged and determined individual, who isn't afraid of good old-fashioned hard work, and doesn't jump ship when the "going gets tough," awaits a profession in the great outdoors filled with opportunities for challenge, adventure, and achievement. But before we examine what the work of the groundie looks like in specific detail (chapters 2-4), let's first take a look at what it looks like in general.

There are at least six prominent features of tree work that you can expect to experience on a regular basis—work that is (1) meaningful, (2) relational, (3) challenging, (4) adventurous, (5) dangerous, and (6) achievement-oriented. Like the ingredients of a good soup, each one will add flavor to your job and put pleasure in your work.

MEANINGFUL WORK

Few things are more motivating than knowing you have something important to do—something worth getting out of bed for. As a ground worker, your work is important—your work matters. Whether you help with tree planting, pruning, repairing, cabling, fertilizing, controlling insect pests and diseases, or removing trees entirely, your work matters because it contributes to the "common good"—the safety, health, and happiness of other people as you make the neighborhoods in which they live more attractive, healthy, and safe. Your work is also important to the welfare of wildlife, as it contributes to the enhancement, preservation, and even the creation of habitat necessary for their survival.

Knowing that your work is important also contributes to your own well-being and happiness. And those who are happy in their work—who have a positive attitude toward it—are more appreciated by their coworkers and bosses, and have much lower rates of conflict, stress, and absenteeism. They also get paid more and get promoted quicker.

RELATIONAL WORK

The moment your employment begins as a tree worker, you become part of a community of people known as arborists, who share a common interest in trees in a profession known as arboriculture. The arborist community is like any other group that gathers around a common interest or belief—churches, schools, branches of the military, athletic teams—in that they share knowledge and skills, advice and wisdom, stories of success and failure, and simply speaking, a passion for the work they do, all to serve the greater good of the group as well as the world at large.

> "Sometimes we just need a friend to let us know we aren't crazy." [Or in some cases—that maybe we are.]
>
> —Drew Hunter
> *(Made for Friendship)*

It's from this fellowship where many arborists acquire much of their knowledge and skill about the trade. It's also where they cultivate some of their most enduring and fulfilling friendships. In a culture such as ours, where friendships can often be shallow, temporary, and unfulfilling, or lacking entirely (perhaps especially among men), there is a great need for faithful friendships and camaraderie with others.

In the arborist community there are many opportunities to find them—at least for those willing to try. These types of friendships are usually first developed with the crew you work with (sometimes even the customers you work for); after all they're the ones who "have your back"—and you theirs—as you toil alongside them day-after-day in the trenches of tree work. However, this "band of brothers" (and sisters) extends well beyond the work place—it is a worldwide community, as it is a worldwide profession.

If you want to see the full display of this community in action simply attend any arborist conference, workshop, or climbing competition. Though you'll certainly notice plenty of activity geared around education and training, events like these are just as much about building friendships—either the forging of new ones or the rekindling of old ones. Granted, friendships can be formed and knowledge dispersed through other means as well, such as social media networking or any number of the on-line arborist forums that are available. But it is through the face-to-face and side-by-side encounters with other tree workers where the most meaningful and lasting friendships grow.

CHALLENGING WORK

Almost every aspect of the ground worker's job is challenging. This is especially true of the first few months of the training phase. You will be challenged *mentally* as you learn what may seem an endless amount of new information and skills; *physically* as you develop the strength and stamina needed to perform them; *emotionally* as you work in hazardous situations and adverse weather; *relationally* as you (sometimes) work amongst crabby coworkers, annoying customers, and critical bosses.

> "At work, we prove and improve ourselves, even find ourselves. If not, at least we earn a living."
>
> —Daniel M. Doriani, *(Work)*

These challenges, however, are not intended to be something you simply endure and *go through,* but rather as something you benefit from and *grow through.* They are the means by which you will grow in knowledge, competence, and character. They will also provide you some of your most meaningful achievements.

Be Teachable

BEING A TEACHABLE EMPLOYEE is the first thing you need to learn to become a successful employee. Surprisingly, *teachability* is a rare quality in people, the lack of which is a huge obstacle to their learning and in getting along with others. We've all encountered unteachable people—those who are wise in their own eyes—the been-there-done-that know-it-alls, who are quick to boast of their accomplishments, state their opinions on almost everything, and make wisecracks and jokes, but slow to listen to sound instruction. For your own sake, and that of your instructors and co-workers: don't be like that! Be teachable.

Being teachable means being willing to learn from others, listening and watching with genuine interest to what's being taught, and using discernment and showing respect when asking questions or offering suggestions of your own. When this happens you are ready to learn—to grow in knowledge and wisdom—and do so more quickly and effectively. And I might add, less painfully—for everyone.

Challenges for Learning New Information & Skills

Acquiring knowledge and skill in tree work will be an ongoing and deeply satisfying part of your job. You will quickly discover that your role as a ground worker is anything but dull, uninteresting, or unimportant. If you think otherwise, look in the list below at the diversity of tasks you may be expected to learn and perform on any given day.

You will learn how to:

- Speak the language of tree work—the terminology of arboriculture.
- Choose, inspect, service, and wear personal protective equipment.
- Choose, inspect, service, and load equipment appropriate for the job.
- Operate and maintain chain saws.
- Inspect and drive work vehicles, trailers, and other related equipment.
- Help with pedestrian and vehicular traffic control.
- Participate in a job briefing.
- Understand a work order and share the instructions with others.

- Identify work site hazards and obstacles and take appropriate steps to protect people and property.
- Talk to customers in a personable and knowledgeable manner.
- Help the tree climber with lowering limbs and managing ropes, clearing the drop zone, sending up tools, etc.
- Install climbing and rigging lines in the tree.
- Correctly tie a variety of knots suited to the tasks at hand.
- Identify trees you work with on a daily basis along with their unique traits.
- Fell trees and/or help with felling trees.
- Limb and buck trees on the ground.
- Lift, carry, and drag heavy loads: tree limbs, trunk wood, and gear.
- Operate a wood chipper, stump grinder, or other related equipment.
- Clean up a job site.
- Respond to emergency situations such as storm damage work.
- Respond to work-related injuries by performing basic first aid or CPR.

On-the-Job Learning

Most of what you'll learn about tree work will occur on-the-job under the direct supervision of a qualified arborist. While people may differ in how they best learn, the general principles of learning are universal. They can be described with the simple phrase: *tell me, show me, let me.* Or, as it's sometimes described: *hear it, see it, do it.* Either way you say it, the process is the same: someone (an instructor) or something (a book or video), *tells* or explains to the student how to perform a task. Next the student is *shown* how to perform the task. And finally, the instructor *lets* the student try the task for him or herself, but still under the watchful eye of the instructor. As the sequence is repeated, the student becomes more proficient.

Interestingly, studies have shown that a student's ability to remember and retain what they have learned is largely dependent on how the information was received. They remember only 10 percent of what they *hear*, but 50 percent of what they *see*, and a whopping 90 percent of what they *do*.[1] So it stands to reason that the most favorable and effective learning occurs when all three of these learning components are used.

A Challenge to Quit

LEARNING HOW TO BECOME a great ground worker is a transformative process; a journey of growth and change in knowledge, skill competence, health and fitness, personal character, and work ethics. But there are also many obstacles that prevent this from happening, the most menacing and lethal of which is substance abuse.

Getting free from alcohol and drug addiction might just be your first (and biggest) challenge as a ground worker; one which promises big rewards if you conquer it. But if you think you can overcome this problem through a half-hearted attempt and an "I can change, if I have to, I guess" attitude, you're wrong.[2] Nothing but all-out warfare on your addiction and an "I want to change because I need to" attitude will do.

About this, arborist/author/speaker, Don Blair pulls no punches, "If you abuse drugs or alcohol, there is no place for you in tree care. Make up your mind."[3] Indeed, and time is of the essence, for what's at stake are peoples' lives–those you work with, those you work for, and of course, your own. Admitting you have a problem is the first step; the second is getting help. And help is available; you don't have to go it alone. If you're not sure where to turn, go to the *Substance Abuse and Mental Health Services Administration* website: **www.samhsa.gov/find-help/national-helpline** or to others like it.

Even if you don't have a substance abuse problem, you may know of coworkers who do. And if their use of alcohol or drugs compromises their work performance or the safety of others, you have a responsibility to make it known to your supervisor. You also have a right and a responsibility to yourself by refusing to work with anyone who is under the influence.

On-Your-Own Learning

Even though most of what you will learn about being a ground worker will come from on-the-job training, don't overlook what you can learn on your own—off-the-job. Self-education is one of the hallmarks of serious and successful students and employees. Serious employees don't wait to be led by the hand to learn what needs to be learned. They take the initiative, fueled by curiosity and enthusiasm, to acquire the necessary knowledge and skills for the job. As a self-educator, your learning will primarily come from relevant books, magazine articles, training videos, arborist equipment catalogs and websites, and especially other arborists. But whether your learning is supervised, self-taught, or both, your goal as a beginning groundie remains the same: to learn as *much as possible* about your work, so that you will become useful and productive as *soon as possible.*

Challenges for Becoming Physically Healthy

Good ground workers need to be physically fit—their work demands it. Ground workers are constantly in motion as their bodies are subjected to repeated movements of pushing, pulling, bending, twisting, and lifting as they cut, limb, drag, buck, and chop limbs and wood. Theirs is a total body workout that combines weight lifting, aerobics and even gymnastics, while often being performed on uneven, littered terrain in adverse weather conditions, using what many consider the most dangerous of power tools—the chain saw. People unaccustomed to the rigors of tree work are a potential threat to themselves and others. They are more susceptible to injuries and fatalities resulting from overexertion or physical and mental exhaustion. While it is certainly true that tree work will get you into shape, it is better yet to get in shape before you even begin.

How you get in shape is beyond the scope of this book. There are countless resources that deal exclusively with that topic. You can begin, though, by making an honest assessment of your current health and lifestyle. Examine yourself in regards to your physical condition, what you eat and drink, what your questionable habits might be (alcohol, tobacco), and how much sleep you get. It may be necessary to make some dramatic changes to your lifestyle for the sake of your work and your coworkers.

Once you are in good physical shape you will most likely maintain that condition simply from working on the job. There is much truth to the arborist bumper sticker that says, "Your yard is our gym." But part-time and seasonal employees need to fill the void of activity in other ways, perhaps through sports activities or by working out at a fitness center.

The customer's property makes a great gym for staying in shape.

Challenges for Building Character

Almost all committed ground workers eventually acquire the necessary knowledge, skills, and physical strength for productive tree work. And as important as those things are, they are not enough. A healthy and competent work crew also needs men and women who have strong inner character—deeply held convictions and values, which are expressed and displayed through outward actions.

Good character is what "separates the wheat from the chaff"—the enthusiastic go-getter from the disinterested slacker, the creative problem solver from the chronic complainer, the patient from the short-tempered, the good-natured from the disagreeable, the steadfast from the quitter. Which would you rather be? Which of these would your boss want as an employee and your coworkers want to work with? The fact is, personal character plays a role that is just as important as knowledge and skill in the overall success of work operations—and life in general. It could even be argued that character is *the* most important asset you bring to your workplace.

Building Character Through the Trials of Tree Work

It has often been said that adversity is what most distinctly reveals character. If this is true then the challenges of tree work provide the perfect school for developing our character as well—humility, integrity, courage, dependability, helpfulness, honesty, patience, perseverance, respect, and self-control for starters. Many factors play a role in this process: the people you work for and with, the outdoor elements you work in, and the very nature of the work itself. Ultimately they can all work together for your good. But again, you'll need to stay on the job long enough to find out.

Building Character by Watching Others

The challenges of tree work are not the only road to building character. Another valuable, but often overlooked path to improvement, is that of watching others. That is, to look for, observe, and then imitate others who already possess good character. For it has been rightly said, "character is better caught than taught." In other words, character is contagious—both good and bad character—so choose your examples wisely. People of good character are everywhere, not without their flaws of course, but still having virtues worthy of imitation. Hopefully, you will find good examples

The Character of William Nindemann

A CREW MEMBER named William Nindemann stepped forward and exerted the full force of his personality [character]. A thirty-year-old immigrant from Rugen Island, Germany, Nindemann was listed as an ordinary seaman, but there was nothing ordinary about him...he was the hardest working, the least complaining, the most resourceful, and the quickest to volunteer for dangerous duty.

The second day out of San Francisco, a ponderous hatchway cover had fallen on his hand and nearly severed one of his fingers. An alarmed Dr. Ambler stitched it up, but the German went about his work without a whimper, refusing to go on the sick list even for a day. "Nindemann is hardworking as a horse," said De Long, and "seems to know no such thing as fatigue."

—Hampton Sides, from *The Kingdom of Ice*

among those you work with. Look also within your circle of friends, family, neighborhood, or place of worship; read about them in books (see sidebar above) and magazine articles, and look for them in films. Building character is a lifelong project though, so be patient, enjoy the process, and most certainly, the results. And as you pursue the path of character, remember: others may be looking at you as an example for them.

ADVENTUROUS WORK

By definition, adventure is an exciting and typically hazardous experience or activity of which the outcome is uncertain. Adventure offers the potential thrill of victory and reward, or the disappointment of defeat. Adventure is also an exploration of self-discovery; to discover the extent of your potential, as well as your limits and prove to yourself, or even to others, that you can achieve something you thought was unlikely or even impossible. Adventure, as described like this, lies at the very heart of production tree work. This, along with a love for being outdoors in the natural world, is a key factor why so many are drawn to the tree care profession. Everyday there is new adventure: new things to learn, new jobs to tackle, new customers to meet, sometimes new coworkers to work with, and always, new challenges and risks that put us to the test to see if we have what it takes.

Fundamentally, the pursuit of adventure is a good and enjoyable thing, especially when it leads to self-improvement and team development. But if we derive pleasure in our work only *because* it is dangerous, we put ourselves and others at risk by glamorizing something that should be treated with a healthy fear and respect. That's why it is so vital to understand just how dangerous tree work can be and adjust your attitude accordingly.

DANGEROUS WORK

While the hazardous nature of tree work may be a significant ingredient for exciting adventure, it is also the primary contributor of injuries and fatalities in the profession. Some of these hazards come from the tools you'll be using and some from the environment in which you'll work. Unfortunately, even the people you work with can sometimes be a threat to job safety.

> "Safe? Who said anything about safe? 'Course it isn't safe. But it's good."
>
> —Apologies to C.S. Lewis[4]

It should come as no surprise then that tree work is always ranked high on the list of most hazardous jobs. The number of ways you can get killed, injured, or experience misery while working is extensive: chain saw injuries, "struck-bys," electrical hazards, slips, trips, and falls, lifting injuries, fatigue, cuts, abrasions, splinters, eye-injuries, poisonous plants, dogs, hornets, bees, spiders, ticks, and snakes.

Being aware of what the hazards of tree work are, how they are caused, and how to avoid them, is the first step in accident prevention (see p. 76 for a more extensive treatment of this topic). The next step is just as important: getting the proper training, pursuing it continually, and putting it into practice every moment of every work day. Ground workers who do this demonstrate that they are serious about the seriousness of tree work.

ACHIEVEMENT-ORIENTED WORK

The need to achieve is a hard-wired desire in most of us. We want to accomplish things. We find great satisfaction in reaching goals and mastering skills that come from hard work. And, when fitting, to receive recognition and reward for those accomplishments. As a ground worker you will have ample opportunities to experience all of these things. Unlike many occupations, where daily accomplishments are often hard to notice, tree workers actually *see* the results of their hard work upon completion of each job, every day. In many cases however, your most noteworthy achievements will take more time to be fully appreciated. Either way, short or long term, you can expect to experience achievement in tree work in one of two ways: as individual achievements and as team achievements.

Individual Achievements

Your first achievements will probably come from the incremental mastery of skills that you formerly struggled so hard to learn: tying knots, "running the ropes," and sharpening chain saws for example. Noteworthy achievements may also come in the form of simply receiving respect and recognition from your employer and crew as they see you make steady improvement. And to the degree that you do, will come greater responsibilities as well. They might come in the form of operating equipment such as a stump grinder or skid-steer loader. Or you might be asked to train and mentor a newly hired ground worker.

As your knowledge and skills increase, possibly to the point of achieving advanced industry credentials, you can also expect increased opportunities for advancement in position, wages, and benefits.[5] While there may be many skilled ground workers who are perfectly content with their position, most prefer to eventually advance to other levels within the profession, such as tree climber, aerial bucket operator, crew leader, tree inspector, or work estimator—who knows, possibly even that of business owner.

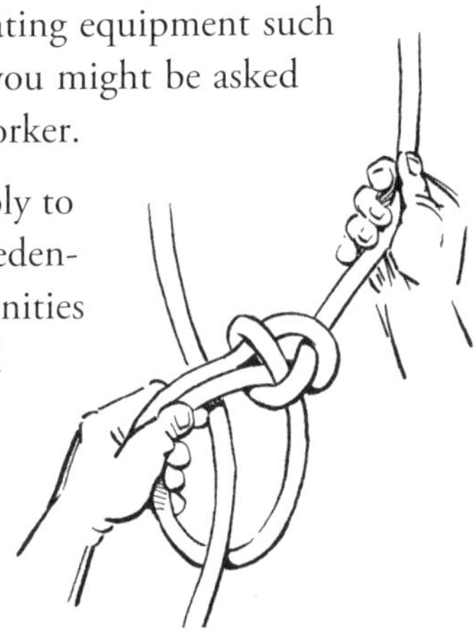

Learning to correctly tie knots will be one of your most satisfying achievements.

Regardless of the work position you choose, the path to getting there always begins as a groundie. Even for arboricultural leaders and educators such as Peter Gerstenberger (Tree Care Industry Association) and Sharon Lilly (International Society of Arboriculture), along with a host of others who climbed to the top of their respective organizations, it all began by first spending time as a ground worker.

Even if you decide on a different type of work entirely, the knowledge and skills you acquire from tree work will last a lifetime. From operating chain saws and tying knots, to working as a team player and conversing with customers—they will prove to be useful in a number of ways in other jobs you may have and during different stages of your life.

Team Achievements

Tree work is a team effort. The way that you and your team function together will determine the outcome of the job. You may end up as one of the most talented members of your crew, but if you can't work with them in a helpful and supportive fashion, the result can be disastrous. Granted, individual performance and achievement matter greatly, as does timely recognition and genuine praise for individual effort. But ultimately, the personal achievements of any worker should somehow contribute toward the greater good of the team and the company you work for.

> **"If someone were to ask me to make a list of the most valuable qualities a person should develop in order to thrive in the world of work—and for that matter, life—I would put being a team player at the top."**
>
> —Patrick Lencioni
> *(The Ideal Team Player)*

Few things in tree work are as energizing and fulfilling than to be part of a team that functions together like a "well oiled machine," working together toward a common goal and accomplishing it safely and skillfully, and doing so in a cooperative manner. To achieve this type of team unity and competence takes time, practice, and a firm resolve not to give up when things get difficult. But it will be worth it. Why? Because the hard-fought victories shared with others will likely become the greatest and most memorable achievements of your work.

PREPARING TO WORK

Now that you've learned what to expect as a ground worker, it's only fitting we end this chapter by describing a few things your employer expects of you. Listed below are a number of preparations you should make even before leaving home or punching in for work. Think of them as your first steps to becoming a great ground worker. To the extent that you practice these pre-work preparations will reveal how committed you are to your job and employer, and to having a safe and productive day of tree work.

- ***Do your homework***—Read your company's employee manual, along with any other related handouts you may be given. Be a self-motivated self-educator by taking the initiative to learn all that you can about the profession in which you now work.
- ***Become fit for work***—Prepare for work by becoming physically and mentally fit. Examine yourself for any questionable habits (drinking, drugs, smoking, chewing, etc.). You may need to make some drastic changes to your lifestyle in order to be a more safe and effective ground worker, and not be a threat to yourself or others.
- ***Rest up, fuel up***—The physicality required of the ground worker is like that of an athlete who competes in strenuous sports. It should be no surprise then that arborists are considered "industrial athletes." Therefore, get plenty of sleep—it will happen naturally if you've been working hard enough—and eat a healthy and substantial breakfast. What you do for your body will determine what it will do for you.
- ***Develop the habit***—To the extent that you can, address "nature's call" *before* punching in for work. Doing so will prevent embarrassing and awkward attempts at the job site and inconveniencing your work crew.
- ***Be prepared***—Bring what you'll need for a day at work. Pack a healthy lunch and beverage (preferably the night before) along with an adequate supply of drinking water. Don't expect to do your food and snack shopping while on the clock, en route to a job. Pack also, any personal items you may need such as an inhaler (asthma), epinephrine auto injector (allergic reactions), or deodorant. Consider also bringing rain gear, a jacket in cool weather, an extra work shirt, or socks to change into after getting sweaty from work or wet from rain.
- ***Show up on time***—Leave for work early enough to get there on time. If you know you're going to be late, contact the office as soon as possible. If you must be absent, request permission as far in advance as possible. Anytime a work crew is shorthanded due to an unexpected tardiness or absence it puts a strain on the rest of the crew, which can have a detrimental effect on work safety and productivity.

- ***Dress appropriately***—Show up for work wearing clothes appropriate for tree work. If your company provides imprinted t-shirts and work shirts—wear them—tucked in if so required. Work pants should be well-fitting, sturdy, and without cuffs (which collect sawdust). Any clothing with holes and tears (which are capable of catching on brush), should be repaired or replaced.

  In general, you should maintain a neat and well-groomed appearance on the job. Remember, you are in the public eye as a representative of your company, so *be* professional in how you look and act. Take pride in your identity as an arborist. For this reason, in addition to safety concerns, jewelry should be removed such as body piercings, earrings, rings, watches, and necklaces. In addition, offensive tattoos should be covered and stickers, patches, and branding on clothes or helmets should be avoided, unless approved by your employer.

- ***Pray***—Seriously . . . pray. In a profession as dangerous and difficult as ours, it would be foolish to ignore such a powerful practice as prayer. But don't wait to pray only in times of urgent need, like when you're on your way to the hospital lying in the back of an ambulance. Pray before troubles come. Pray in the morning before going to work. Ask for protection, for skill, for endurance, for wisdom, for patience. And pray not only for yourself, but also for your crew, your boss, and even your customers. Amen.

THE JOURNEY BEGINS

Congratulations! You've made it to the end of the first chapter. For some of you, this in itself might be a significant achievement (finishing the book an even greater one). But now the real journey and challenge begins, as you head out to work and begin learning and doing what ground workers do—***at the shop*** (the topic of the next chapter), ***on the road*** (chapter 3), and ***at the job site*** (chapter 4).

2

The Role of the Ground Worker...

At the Shop

"The shop"—or yard, or garage, or whatever it is you call the place where you meet in the morning—is where plans and preparations are made for the days work. This is where the work order is discussed and preliminary work plans formed and put into action; where work teams are selected and instructed where to go and what to do; where equipment is chosen, maintained, and loaded on to work vehicles. It's here at the shop where success at the job site all begins.

UPON ARRIVAL

Preparations for the work day ahead begin the moment you arrive at the shop and punch in for work. For starters, begin your work by thinking about and talking about work, rather than the activities of your weekend or the night before. There will be plenty of time for sharing personal stories on the job, but for now you need to prepare your mind for action.

If the company you work for has an established "morning ritual"—a set of work procedures routinely carried out each morning—then the next thing to do is to simply get busy carrying out those tasks. In many cases employers require their workers to first check in with their boss or crew leader for work instructions. In either case, if you are ever at a loss for something to do, simply ask someone who is "in the know" for directions. With so much to do, there is no excuse for walking around with your hands in your pockets.

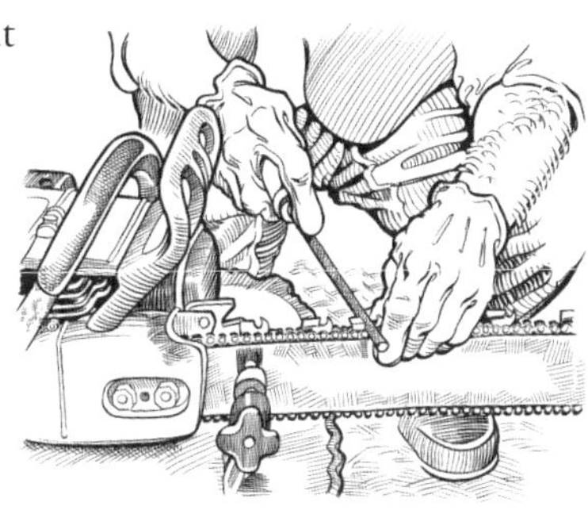

Preparations made at the shop help determine success in the field.

The First Five Minutes

THE FIRST FIVE MINUTES of your morning exchange with boss and coworkers may just prove to be the most important minutes of your day as well as your coworkers'. Studies have shown that moods and emotions, both positive and negative, whether expressed through words, tone of voice, expressions, or body language, are highly contagious.

A grumpy, cheerless, or indifferent response to a coworker's morning greeting can sour the relationship for hours—possibly the entire day if the negativity persists. Of course the opposite is also true—a cheerful, positive, engaged greeting is more likely to bring out the same response from another and thus set the stage for a potentially pleasant and productive day working together.

Interestingly, studies have shown that having just one happy player—or worker—on a team improved not only team morale but also the results. Make it your goal then, to be one of those happy—or at least pleasant—kind of workers, who has a positive influence on those they work with. Doing so could be one of your most impactful contributions in the day of another—it all depends, of course, on the first five minutes.

THE MORNING MEETING

Aside from the morning ritual tasks, most companies also hold some type of a job briefing, or morning meeting, before heading out to the job site. It is from this meeting that you will learn such things as who the customer is, how to get to their property, what type of work will be performed, and what tools and equipment will be needed to most safely and efficiently execute the work. Later on, after arriving at the job site, you will be briefed again about the work order and work plan, but this time with more specific details (more on that in chapter 4).

As a ground worker, you may or may not be included in the morning meeting held at the shop. If you are not included, you will be informed of all relevant details at the appropriate time. If you are included, then pay careful attention to what's being shared, but at the same time don't be afraid to ask relevant questions. Don't be one of those people who is afraid to ask questions because they don't want to appear ignorant. Asking questions is the mark of a serious student.

GETTING GEARED UP

As a beginning ground worker you will quickly discover that there are a wide variety of tools and gear used in tree work. Each item has a specific and sometimes multi-purpose function. But if you don't know what they're used for, or where to find them, or how to maintain, repair and operate them, you won't be of much use either at the shop or in the field. Therefore, one of your first assignments as a ground worker is to learn the tools of your trade—those necessary to perform the work (production tools) and those necessary to keep you safe while you work (personal protective equipment).

Gettin'er Done: Production Tools

The sooner you find out what a biner, block, ascender, load line, throwline, throwbag, Port-a-Wrap, Big Shot, whoopie sling, cant hook, or come-along is, the better. These are but a small sampling of the oddly named tools you're likely to use on a daily basis. At the very least, these are tools you'll be asked to locate at the shop or fetch from the work truck. These tools, along with a number of others, are commonly referred to as production tools—tools for getting the job done safely and efficiently.

The back-saving cant hook tool.

Most of these tools are described, illustrated, or referred to at some point in this book (see glossary on page 174). The ones which are not will be learned while working on-the-job. And for the motivated self-educated type of tree worker, there are many resources from which to learn about arborist equipment (see page 173 for suggestions). You can also learn a wealth of information about arborist equipment by simply paging through an arborist supply catalog or visiting their websites.

Stayin' Alive: Personal Protective Equipment (PPE)

Personal protective equipment, or simply PPE, is the most important, and thus necessary, equipment you need to obtain from your employer and learn how to wear, inspect, and care for properly. This is the equipment that is designed to protect your body and most importantly, your life.

As it relates to PPE, tree workers and their employers have a shared responsibility to comply with applicable regulations, policies, and standards established by the Occupational Safety and Health Administration (OSHA) and the American National Standards Institute (ANSI). Employers are required by both OSHA and ANSI to provide their employees with appropriate PPE and instruction in its proper use and care. Should you choose to purchase your own PPE, do so only after consulting your employer and work supervisor to ensure you get the correct equipment. Let's take a closer look at each item of PPE.

Head Protection

When you consider that the leading cause of death and injury for ground workers is from falling objects from above—known as "struck-bys"—it makes perfect sense that the safety helmet is the most important piece of safety equipment they can own, but only if it's worn constantly and maintained properly. There are a number of styles to choose from, as can be seen from looking at any arborist equipment catalog. But regardless which one you choose or that your employer provides, it must meet the requirements of ANSI Z89.1 for general duty work for arborists.[6]

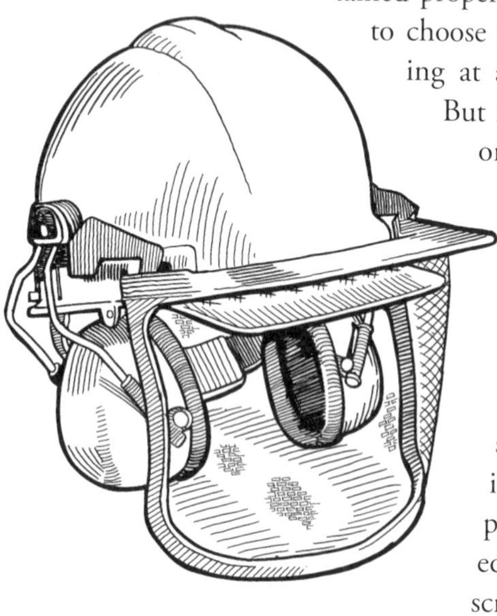

The ground worker's best friend.

One approved helmet style that is particularly popular with ground workers is the "helmet system." This type of helmet integrates head, face, and ear protection with one piece of equipment. It has an attached screened (or solid plastic) face shield which flips up easily when not needed or back down when the work gets messy and more dangerous. Similarly, the attached ear muffs can be raised onto the helmet during periods of noise inactivity or down when operating noisy equipment.

ANSI Z133: The Last Word on Safety

TO WORK SAFE and be safe is the overarching goal of the tree worker. It stands to reason then, that consulting *ANSI Z133—Safety Requirements for Arboricultural Operations* is a wise first step toward meeting that goal. This is a publication of safety standards for tree care operations published by the International Society of Arboriculture or ISA. ANSI Z133 is the recognized safety standard for tree care operations in the United States (tree workers in Canada comply with the standards set forth by the Canadian Standards Association, or CSA). These safety standards are developed and revised on a regular basis (traditionally about every five or six years) by a committee made up mostly of tree care professionals.

Although the publication refers to the standards as requirements, compliance is voluntary. Nevertheless, the standards presented in the ANSI Z133 publication are the industry consensus for safety and should therefore be learned and practiced by all tree workers. The safety requirements discussed in this book are derived from the ANSI standards. The words ***should*** and ***shall*** are important and consistent words that appear in the ANSI standards. *Should* refers to an advisory recommendation whereas *shall* refers to a mandatory requirement.

Reference to ANSI is included in this book on a limited basis. Therefore, wise and responsible employers ***shall*** obtain a copy of the ANSI Z133-2017 publication (the most current edition at the time of this printing) and make sure their employees are instructed on all the requirements relevant to their work situations and follow them faithfully.

Orange is perhaps the most popular color, especially for a ground worker, as it increases your visibility to other workers, most importantly of whom is the climber above. You should perform a close inspection of your helmet at least once a week for cracks, frayed straps, or any other signs of wear or damage to the shell, suspension, or helmet components. Repair of the helmet itself is rarely an option. Damaged or worn helmets should be retired, destroyed (so no one else wears it), and replaced. Even helmets not showing any noticeable damage or wear should be replaced routinely—every four or five years is not too frequent for an inexpensive piece of life saving gear such as this.

Eye Protection

Getting poked in the eye with a branch will stop you dead in your tracks. Even a small wood chip, fine sawdust, or metal saw chain filings on the eye surface can be momentarily crippling. Wearing eye protection will significantly reduce this from happening and protect from other types of eye injuries. With so many cool designs available these days there's hardly an excuse for not wearing eye protection. Besides, along with head protection, ANSI Z133 requires their use "when engaged in arboricultural operations."

Safety glasses must also comply with ANSI Z87.1 specifications; they will be stamped to indicate that they do. The cheap glasses sold at gas stations and discount stores just won't do it. For workers who wear prescription eye glasses, you might consider using helmet systems with a solid surface face shield (which also meets Z87.1 specifications), or purchasing prescription safety glasses, or using a combination of contact lenses with standard safety glasses.

Face Protection

The screened face shields mounted on the helmet systems previously described offer excellent face protection as they stop or deflect small tree limbs, flying debris, and wood chips. While they do offer some eye protection, face screens are not a substitute for protective glasses. Fine sawdust can easily pass through the screen openings and wood or metal projectiles can still manage to rip through either the metal screen or plastic mesh. Screens are inexpensive, so replace them as soon as rust, tears, frays, or holes develop on the surface. If solid face shields are being used (which again, meet Z87.1 specifications) it is not necessary to wear safety glasses.

Hearing Protection

Many veteran tree workers who were exposed to the constant racket of chain saws and wood chippers wish they had protected their hearing when they first started their tree work careers. Take my advice, as someone who has "been there, done that," and as a result wears hearing aids: *protect your hearing!*

Two effective ways to protect your hearing are ear plugs and ear muffs. Plugs are inexpensive, compact, and disposable, but can be awkward to install, especially with gloved hands, and they quickly get dirty. Muffs are reusable, and if mounted on your safety helmet, quick to employ and convenient to store. They also help keep your ears warm in cold conditions, which of course means they can be warm in hot conditions.

Another less frequently considered benefit of muffs is the protection they offer to the ear itself. Anyone who has been whacked in the ear or suffered a punctured ear drum from wrestling through thick brush and downed trees can appreciate the added protection muffs provide. Be aware that safety glasses can inhibit a proper seal when worn with muffs, thereby compromising the muffs' effectiveness. You might consider wearing both plugs and muffs for maximum protection.

Hand Protection

Work gloves basically do two things: protect your hands from injuries such as cuts, scratches, splinters, and burns and provide a firm grip with less effort on the things you handle daily while working—limbs, logs, chain saws, rope, and other work related tools. Choices include traditional leather for general work, latex-dipped "gripper gloves" for line handling and tool gripping, and styles that are chain saw resistant. The wide-cuffed "gauntlet" style gloves, commonly sold at hardware stores, are not suited for tree work, as they fill with sawdust and catch easily on brush, which can be especially dangerous when feeding a wood chipper.

Leg Protection

Since the majority of chain saw injuries occur on the legs and knees it should seem obvious that wearing protective pants, chaps, or bibs designed to protect them is a wise thing to do. Besides, ANSI and OSHA require their use, as should your employer. Regardless of the style worn, none are cut-proof. Instead, the fabric of the leg protection is designed to slow or jam the cutters of the saw when contact is made, thus reducing the chance or severity of injury.

Foot Protection

There are many features that make for a good work boot. Select a pair that will provide adequate arch and ankle support, good traction in a variety of terrain and climate conditions, insulation against cold weather or water, and a measure of protection in the toe area.

Regarding toe protection: to wear or not to wear boots with safety toes (steel-toed, etc.) is a controversial topic amongst tree workers. Many strongly advise against them, while others swear by them. So, before making a purchase do some research on arborist footwear and always check with your employer first. That said, if you do decide against boots with safety toes, then make sure they are heavy and durable enough to offer some measure of protection against chain saws or dropped log ends.

While some low-cut boots may meet the above criteria, they tend to collect sawdust and therefore should probably be avoided. Realistically, no boot can do it all. Experienced tree workers know this and eventually end up owning several pairs of work boots to meet the diverse work, weather, and terrain conditions involved in tree work.

Work Clothing

As was mentioned earlier, if your employer furnishes work shirts and/or pants, then make sure you wear them and care for them properly. If not, then just about any well-fitting sturdy work pants and shirts without holes and tears (which can catch brush) are acceptable for work. Avoid wearing pants with cuffs, which collect sawdust, and hammer loops, which can potentially snag brush. As for shirts, long-sleeved offer the best protection for the arms against scratches and insect bites, but can be warm during some parts of the year. As for jewelry—leave it at home or in the work truck.

A well dressed and well protected ground worker.

Tree workers who frequently work near roadways may also be advised or required to wear high-visibility apparel to help prevent being struck by vehicular traffic. Fluorescent yellow-green and orange-red are the most effective colors for this purpose.

Taking Care of the Gear

Not only is it important to bring all the necessary equipment to the job site, it must also be in good condition and function properly. Compromised equipment leads to compromised work safety and productivity. Therefore, inspecting and maintaining equipment is of the utmost importance. Even though this is an ongoing activity throughout the day, it is one that first begins at the shop in the morning—an activity that ends there as well. It is here where you have the first opportunity to notice damaged equipment and have it repaired or replaced before heading out to the job.

As equipment is being loaded onto the work vehicle or removed temporarily to service (as when fueling and sharpening saws), workers should pay attention to any equipment that appears to be damaged and defective or shows sign of excessive wear and deterioration. Ground workers should be particularly attentive to the equipment they personally use: rigging equipment such as ropes and hardware, chain saws, and equipment used for job site clean up. If any equipment damage is discovered it should be reported to your supervisor or crew leader who will determine whether it should be removed from service and replaced, or repaired and put back into operation.

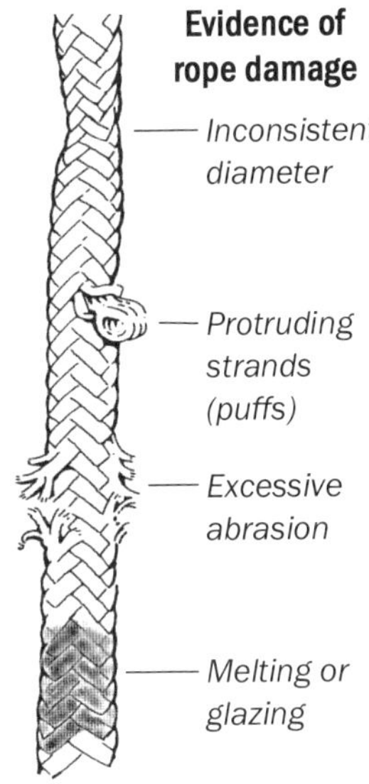

Evidence of rope damage

Most of the morning maintenance tasks require little time to perform—fueling chain saws, touching up saw chains, replacing a broken rake handle, heat sealing or taping frayed rope ends, restocking the first aid kit, or simply filling the water jug. However, more elaborate tasks such as changing chipper knives or servicing work vehicles will require significantly more time to perform. And the longer it takes to service and repair equipment at the shop will be that much longer before you can get on the road. That's one reason why some employers prefer to have equipment serviced at the end of the work day rather than in the morning—time permitting of course. This is particularly important during storm work when crews need to be mobilized as quickly as possible.

Keep in mind also, that it is almost always more efficient to maintain and repair equipment at the shop than out in the field—though you will need to do that too. Not only are the best tools found there, along with the most competent person for performing those tasks, but an enclosed shop also affords a comfortable place to work out of the weather.

It goes without saying perhaps, but many repairs made to equipment could be avoided in the first place if more care was taken by those who use them—if they treated the equipment as if it were their own. Consider applying the "Golden Rule" to your company's equipment: "Whatever you wish that others would do to your equipment, do also to theirs."[7]

MAINTAINING THE CHAIN SAW

The chain saw is the workhorse of tools for ground operations in tree work. But it is useful only to the extent that it is operated safely and maintained properly. In the same way that skilled craftsmen know their tools—how to use them, inspect them, clean and service them, even love and respect them—so should you know how to operate and care for the chain saws you use. For this reason, significant attention is spent in this chapter on chain saw maintenance. Chain saw operation and safety is covered in chapter 4.

> "A poor cutting saw is of no advantage to anyone, particularly for the beginner who can't distinguish the difference between a saw that cuts so-so and one that cuts well."
>
> –Gerald F. Beranek *(The Fundamentals of General Tree Work)*

Again, an enclosed shop provides the ideal environment for working on chain saws—a bench vise for stabilizing the saw, an air compressor when it needs a thorough cleaning, and a roof to stay dry and comfortable. But most, if not all maintenance tasks, can be and will be, performed just as frequently in the field—especially sharpening saw chain.

But either way, in the field or at the shop, the same basic inspection and maintenance principles apply and the same basic tools are required to perform them. Before we look at saw maintenance procedures it will be necessary to first obtain the proper tools.

Tools for Chain Saw Maintenance

The list of maintenance tools listed below may seem extensive, but with the exception of the last two items, they are small enough to fit into a small tool box or pouch. In most cases, however, field maintenance will only require the use of the appropriate round file, a flat file, and a "scrench" (wrench and screwdriver in one tool). The list includes:

> **"Tools in their own right are exceptionally pleasing objects."**
>
> –Max Bainbridge (spoon carver)

- Chain saw owner's manual
- Eye protection when cleaning and sharpening the saw; head and hearing protection when running one
- Snug-fitting leather gloves (especially when sharpening chain)
- Stiff bristle brush for general cleaning (a tooth brush works great)
- Round files with handle (for cutters)
- Flat file with handle (for taking down the depth gauge/raker and dressing the guide bar rails)
- Saw chain filing guide (optional) and depth gauge tool
- Wrenches to fit the saw (spark plug, sprocket cover lug nuts, etc.)
- Screwdrivers to fit saw (saw chain tensioning screw, carburetor adjustment screws, air filter cover, screws and bolts on saw body, etc.)
- Multipurpose tools such as a "scrench" or the TopSaw® 12-in-1 tool, which combines wrench and screw driver(s) in one tool
- Clean out tool for guide bar groove
- Small piece of hooked wire/hooking tool (fuel and oil filters)
- Chalk, crayon, or marker pen (to mark the "master cutter")
- Cleaning rags
- Tool box or pouch for storing tools
- White lithium grease (for guide bars that require it)
- Stump vise (field) or bench vise (shop)
- Air compressor (for thorough cleaning)

Chain Saw Maintenance Procedures

What follows is a description of general maintenance procedures for all chain saws. Consult the owners manual specific to your chain saw if you need a more detailed description of service recommendations and procedures. The following procedures can be performed in the field as well as the shop. Most will be performed daily, or as needed; others only weekly.

Since the procedures that follow can be performed without the saw running, the only PPE needed is eye protection and gloves. If it is necessary to run the saw, learn how to properly start it by referring to pages 119-120 and wear other appropriate PPE, such as a helmet, and possibly even leg protection. Beginning ground workers should become familiar with chain saw anatomy presented on the opposite page and saw chain anatomy on page 41 before performing any saw maintenance for the first time, and even then they should do so under the supervision of a qualified worker until they become proficient.

Chain Saw Body

- ☐ Make a visual inspection of the overall condition of the saw body; look for cracks, and gas and oil leaks. Repair or replace parts as needed.
- ☐ Remove any excessive saw dust, dirt, and debris with a rag or soft brush. Remove pine sap/pitch from front and rear handles using gasoline, paint thinner, or mineral spirits.
- ☐ Check for and tighten any loose nuts, bolts, and screws. This is especially important within the first 100 hours of new chain saws.
- ☐ Inspect bumper spikes: tighten, straighten, repair, or replace.
- ☐ Inspect to ensure throttle trigger interlock works properly.
- ☐ Test/activate the chain brake to ensure it is functioning smoothly (with the engine off).
- ☐ Make sure the chain catcher is secure and undamaged.

Air Filter

- ☐ Remove the air filter cover. Before removing the air filter itself, activate the choke lever to prevent debris from entering the carburetor during cleaning.

Chain Saw Anatomy

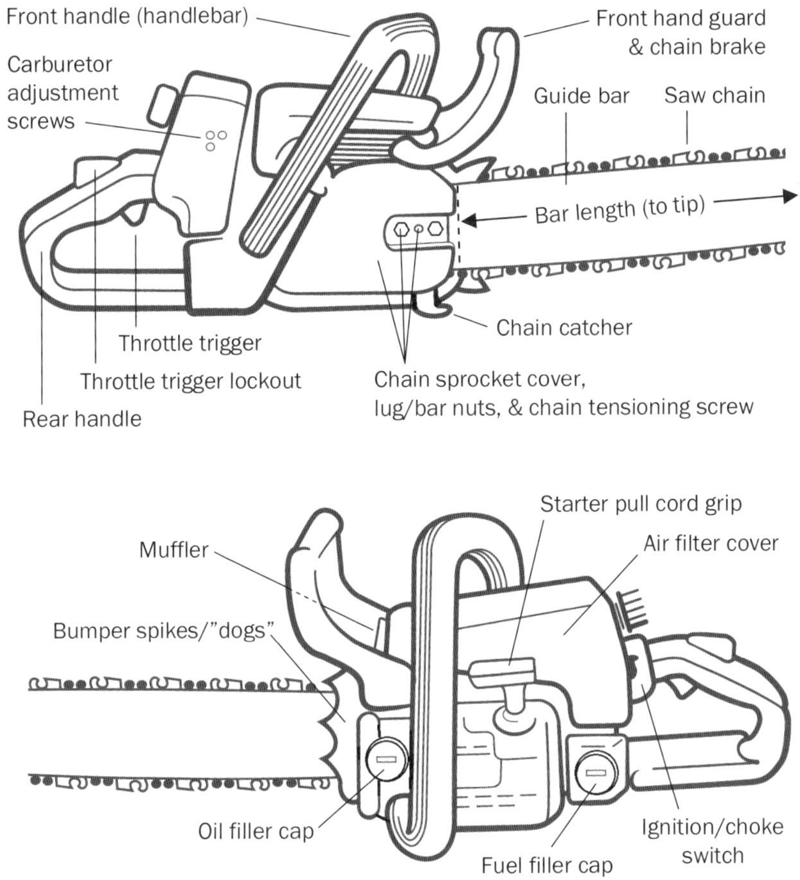

☐ Remove the filter and clean, or replace if damaged. Use the palm of your hand, or a hard flat surface, to knock off the bulk of debris, followed by a gentle brushing with a toothbrush when field cleaning, or compressed air when at the shop. When using the latter method, always blow from the inside of the filter outwards. Never use gasoline to clean the filter.

☐ Before installing the cleaned filter and filter cover, apply the choke and use low-pressure air to blow out the area around the carburetor.

Starter Pull Cord & Starter Pulley

- ☐ Inspect daily the starter pull cord for frays and any other sign of wear. Replace if necessary.
- ☐ Be sure the pull cord is tensioned properly and the correct length. Consult the owners manual to properly tension the starter pulley or to replace the pull cord.

Guide Bar

- ☐ Remove the chain saw sprocket cover to remove the guide bar and saw chain from the saw. Place the lug nuts inside the cover to prevent loss.
- ☐ Visually inspect the saw bar, or ***guide bar body***, to ensure that it is straight and not bent by sighting down the length of the bar. Use a small square, or other reliable straightedge, against the face of the bar to check for more accuracy. Repair or replace as needed.
- ☐ Inspect the ***guide bar rails*** for minor nicks, burrs, and defects. Use a flat file to remove and repair. Check also that the rails are straight, flat on top, and not pinched or splayed.
- ☐ Clean the entire length of the ***guide bar groove*** using high-pressure air, or a thin screw driver, pocket knife, or a guide bar cleaning tool to remove any metal filings, sawdust, and imbedded bar oil. (I learned from a rookie groundie that an old credit card, or the like, also makes a good groove cleaning tool in a pinch.) Make sure the oil hole is clear as well.
- ☐ Inspect the ***guide bar nose sprocket*** to make sure it spins freely.
- ☐ Scrape off any oil/sawdust buildup from the end of the bar that attaches to the saw to ensure a good mounting seal.
- ☐ It is important to remember to "flip the bar"—that is, turn the bar over—before reassembling. Doing so will help prevent one-sided wear on the underside of the bar and at the nose.
- ☐ Lastly, clean the inside of the sprocket cover and the entire area on the saw body under the cover with a scraper or screwdriver, rag, and, when available, compressed air. This is also a good time to inspect the chain sprocket for wear and replace it if necessary. Replace the chain saw sprocket cover and only slightly tighten the lug nuts so that the chain can be tensioned according to instructions on the next page.

Tensioning the Saw Chain

- ☐ With the lug nuts only slightly tightened, or loosened if they were previously tightened, lift up on the bar nose with one hand and use the other to turn the tensioning screw clockwise to tighten a loose chain or counterclockwise to loosen a tight chain. If the saw is secured in a vise, lift up on the saw handle rather than the bar nose. A properly tensioned chain should fit snugly against the bottom of the guide bar yet loose enough that you can turn the chain by hand.
- ☐ Once proper chain tension has been attained and *while still holding the bar nose up*, alternately tighten the lug nuts, starting with the rear one.

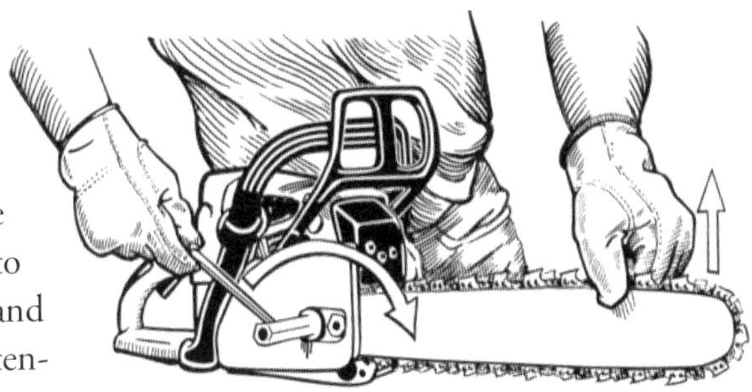

Lift up the bar nose as you tighten the lug nuts.

Note: It is recommended that you wear gloves whenever you check the chain tension—one slip while trying to turn the chain by hand (especially a tightly tensioned one) is likely to lead to cut fingers and a search for band-aids. Also, it is important to check the chain tension periodically. This is especially important with a new chain which will stretch more readily than one that is broke in. Running the chain too loose can cause the saw to cut erratically and the chain to derail—to jump off the bar. Running the chain too tight can cause excessive bar and chain wear and heating, which can lead to the chain stretching.

Miscellaneous Maintenance

- ☐ Spark arrestor screen: inspect and clean.
- ☐ Spark plug: remove, inspect, clean, adjust, or replace.
- ☐ Carburetor adjustments: making these adjustments requires skill and training and therefore should be done by those who are qualified to do so. However, ground workers should at least know how to adjust the idle speed of the saw. This is easy to learn and extremely important in preventing the chain from moving while the saw is idling.

Fueling the Saw

- ☐ Lay the saw down on an appropriate surface that is flat and stable. Never fuel saws on a customer's lawn, driveway, or sidewalk unless a tarp or other protective ground cover is used. Spilled gas and bar oil can cause permanent damage to any of these surfaces.
- ☐ Before fueling the saw, clean the fuel cap and the area around it with a rag or brush to prevent dirt and debris from entering the tank and contaminating the fuel.
- ☐ Identify and remove the fuel cap using a scrench or screwdriver, or your hand with saws equipped with toolless caps. Remove cap slowly to gradually release any pressure build-up in the tank.
- ☐ Gently shake the fuel/oil mixture and pour into the fuel tank—not the oil tank! Learn to identify which is which. Use a funnel or flexible nozzle. Do not overfill; leave room for gas expansion in the tank. ***Never smoke when fueling a saw!***

  Note: If you mistakenly put bar oil in the gas tank—no harm done. Simply dump the oil back in the jug, fill the gas tank half way with gas, shake it, and dispose of properly. Two times should do it. As for putting fuel in the bar oil tank? Just dump out the gas and fill with bar oil.
- ☐ Replace the fuel cap and tighten securely—but do not overtighten. Wipe up any spilled gas on the saw or fueling surface.

Filling the Bar Oil Tank

- ☐ You should fill the bar oil tank each time you fuel the saw using the same procedure for fueling saws: clean the oil cap and surrounding area before removing the cap; remove the cap with a scrench, screwdriver, or your hand; carefully add the correct bar oil—winter or summer weight—and do so without spilling, or accidentally pouring into the fuel tank; securely replace the caps and clean up afterward.
- ☐ It is important to learn just how tight the fuel and oil caps should be secured. Caps which have been tightened too much are frustratingly difficult to open and can be damaged while attempting to wrench them free; and loose caps invariably work themselves open spilling their contents on the nearest target—usually you, the saw operator.

MAINTAINING THE SAW CHAIN

Operating a chain saw with a dull or improperly sharpened saw chain is a dangerous activity, and practically speaking—a waste of time. It increases the chance of kickback (see page 123), contributes to premature wear on the chain sprocket, guide bar, and the chain itself, and wastes time and money as cutting efficiency is diminished. What's more, a poorly cutting chain saw is simply not any fun to use. Reasons enough to keep it sharp with well-timed sharpenings—preferably done before the "heat of the battle" of a challenging tree job.

> **"When the battle drum beats, it's too late to sharpen your sword."**
>
> —Winston Churchill

When It's Time to Sharpen or Replace

1. When the chain *looks* dull or damaged; top or side plate cutting edges are visibly damaged, nicked, or rounded, the cutting tooth or depth gauge is broken off, or the chain rivets are damaged. Make it a habit to inspect the chain often and always before leaving the shop in the morning.
2. When significant pressure needs to be exerted to the bar to get the chain to cut wood. A properly sharpened chain should cut through wood with ease and with little pressure exerted on the bar.
3. When the saw produces fine sawdust rather than coarse wood chips.
4. When smoke comes out of the saw cut, but not the kind that is caused from inadequate chain lubrication or improper chain tension.
5. When the saw cuts are angled, wandering in one direction or the other. This is an indication that the cutters are of uneven length or that they are dull on just one side of the chain.
6. When the saw jumps or chatters during cutting (an indication the depth gauge/raker is too low), or excessive pressure is needed to cut the wood (an indication the depth gauge is too high).

Of course, one of the best ways to avoid having to sharpen saw chain in the first place, or at least so frequently, is by paying attention to and avoiding the things that cause them damage—nails, concrete, rocks, and dirt. Nevertheless, even the best cared-for chains get dull eventually and thus require proper sharpening to regain their cutting efficiency.

Replacing the Chain

Even though a chain may be cutting poorly, it may not be worth sharpening. If the cutters are considerably worn, damaged, or used up, then it's time to swap it out with a new chain—or a sharpened used one. Another instance when it's more sensible to replace a dull chain rather than sharpen it, is in the field when there isn't time to sharpen it properly.

In either case you will need to know three important numbers, or measurements, to ensure that the replacement chain will be the correct size for the chain saw and guide bar being used: the *pitch measurement*, the *gauge measurement*, and the *number of drive links* on the chain. Knowing these numbers is particularly important when purchasing chains.

Pitch—The pitch of a saw chain is the distance between any 3 adjacent rivets on the chain divided by two. One of the most common pitch sizes is 3/8 inch—a distance of 3/4" between 3 chain rivets, divided by 2.

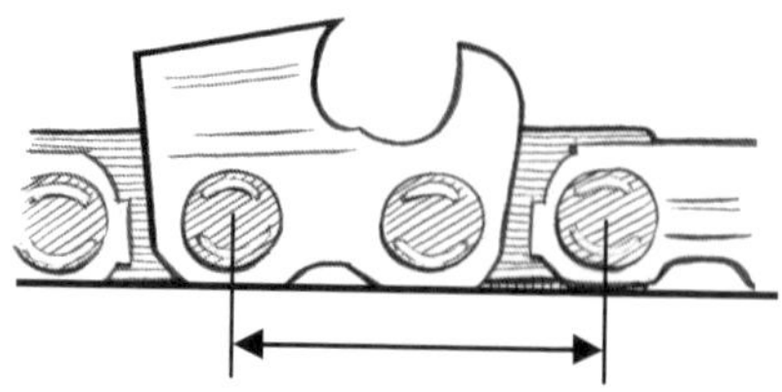

Distance between 3 rivets ÷ 2 = chain pitch.

Gauge—The gauge measurement, in thousandths of an inch, refers to the thickness of the drive link, that is, the part of the chain that fits into the guide bar groove. The most common gauge measurements for saw chain are: .043", .050", .058", and .063".

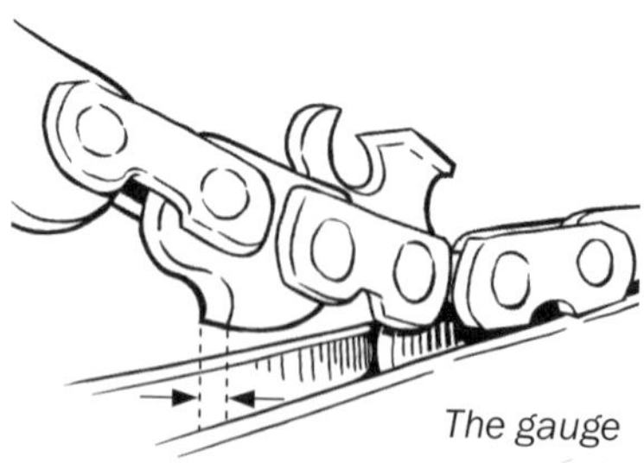

The gauge measurement.

Note—Most saw chain manufacturers have a code number stamped either on the drive link, cutter tooth, or both, which corresponds with the chain's pitch and gauge measurement (see illustration on page 41). For example, drive links stamped with the number 72 signifies a 3/8 pitch, .050" gauge chain. Knowing these details makes chain replacement much easier.

Number of drive links—To match chain length to the length of the guide bar you must know the number of drive links on the chain. If this number is not already known, it will be necessary to count the links individually.

Saw Chain Anatomy

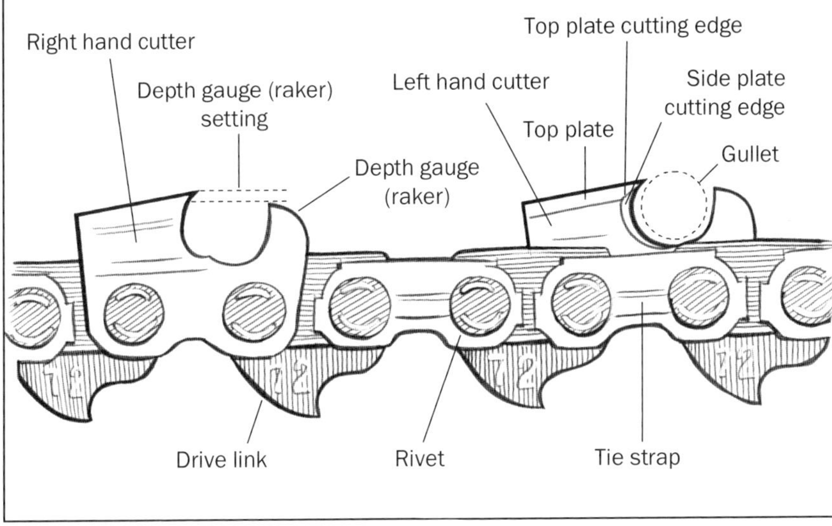

Sharpening the Chain

Sharpening a saw chain is one of the most important skills a tree worker can learn, which is why so much attention is devoted to it in this chapter. Therefore, it's important to learn it correctly under the guidance of someone who knows what they are doing. It has often been said that "practice makes perfect," but actually, "practice makes permanent"—and sometimes perfect. In other words: bad habits are hard to break; learn correctly from the start. The sharpening procedures, principles, and tips that follow will get you started in the right direction, but you'll still need hands-on instruction from an experienced sawyer, along with lots of practice.[8]

Sharpening Options

The two main methods for sharpening saw chain are with a bench mounted grinder or by hand using a round file for the cutters and flat file for the depth gauges (rakers). Motorized sharpening in the shop certainly has its place, but since much of your saw chain maintenance will be performed in the field with the chain on the saw it is imperative for the ground worker to learn manual chain sharpening. Besides, manual sharpening is considered by most to be the cheapest, quickest, and all around best method available.

Sharpening Preparations

Before you begin sharpening the chain, and filing the depth gauges (rakers) later on, perform the following steps of sharpening preparation:

1. Become familiar with the saw chain terms and anatomy (page 41).
2. Wear snug-fitting gloves to prevent hand injury.
3. Wear safety glasses to protect your eyes against dust and metal shavings.
4. Select the proper sized round file for the chain. This is determined by the chain pitch (page 40). Some of the more commonly used chains, along with their recommended round file sizes, are as follows:
   - 1/4" pitch = 5/32" (4.0mm) round file
   - .325" pitch = 3/16" (4.8mm) round file
   - 3/8" pitch = 13/64" (5.2mm) or 7/32" (5.5mm) round file (chain manufacturers vary in file size recommendation). Some workers claim having best results by starting out with the larger 7/32" sized file and switching to the slightly smaller 13/64" sized file as the cutters get more worn down
   - 3/8" pitch low profile (Picco) = 5/32" (4.0mm) round file
   - .404" pitch = 7/32" (5.5mm) round file
6. Finally, gather up any other tools you'll need for sharpening, such as a flat file, scrench, stiff brush for cleaning, etc. You're now ready to sharpen the chain.

Sharpening Procedure (Part 1): Filing the Cutters

1. ***Secure the saw***—Whenever practical secure the chain saw in a vise while sharpening—a bench vise at the shop, one mounted on the tailgate of a work truck, or a portable one driven into the end of a stump (unsurprisingly called a stump vise—see page 23). If no vise is available, get creative. It's usually enough to simply hold down the tip of the guide bar against a flat wooden surface, such as a log, stump, or sheet of plywood mounted on a tailgate of a truck.
2. ***Tighten the chain***—The chain should be tensioned tight enough to prevent the cutters from tipping during sharpening. This may mean tightening it more than normal and then loosening it afterward.

3. ***Clean the saw chain***—Use a small bristle brush to remove any residual sawdust or debris from the chain. It may be necessary to use a cleaning solvent—or even chain saw gas in a pinch—and wire brush to remove any excessive sap buildup on the cutters. This is also a good time to make note of any chain damage.
4. ***Find and mark the "master cutter"***—Locate the shortest or most damaged cutter and mark the top plate or tie strap with chalk, a crayon, or felt-tipped marker. This "master cutter" will be your starting point. Ideally, all other cutters should be filed back to the same length as the master cutter, but only when it is practical to do so.
5. ***Determine the sharpening angle***—On most chains the recommended top plate sharpening angle is imprinted on the top surface of the cutter (fig. 1). This "service mark" is typically either a 30 or 35-degree angle (30 being most common). Position the file so it is parallel with this mark.

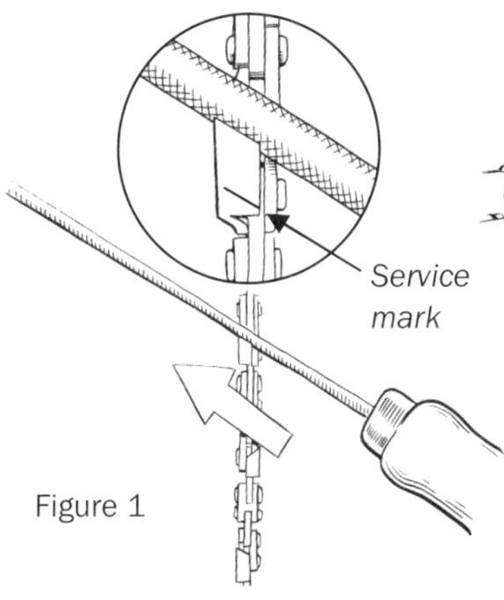

Figure 1

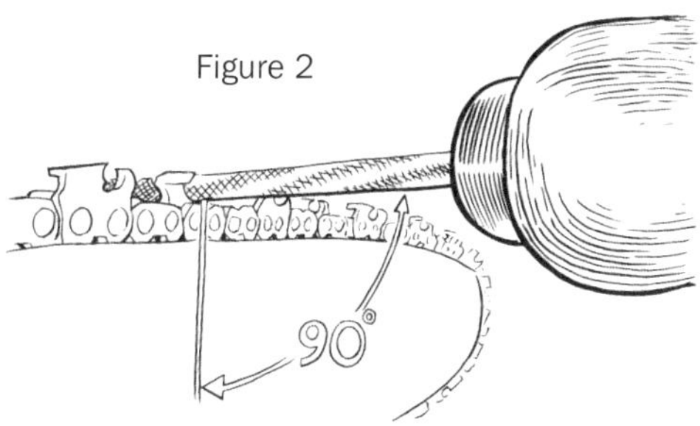

Figure 2

Position the file so it is parallel to the service mark (fig. 1) and 90° to the flat surface of the guide bar (fig. 2).

6. ***Sharpen the cutters***—Beginning with the "master cutter," hold the file 90° to the flat surface of the guide bar (fig. 2) and at the proper top plate sharpening angle (fig. 1). Apply forward pressure on the file with one hand, while guiding it with the second, to maintain the correct sharpening angles. If the bar is not stabilized in a vice, you might only be able to use one hand to guide the file, if the other hand steadies the bar. In either case, activating the chain brake will help hold the chain in place. Always file the cutter from the inside outward. Lift the file slightly on the return stroke to prevent file wear and cutter damage.

Sharpen the cutters (continued)—File the cutter until all the nicks or chips are removed and the cutting edge is straight and sharp. This shouldn't take more than 5-6 strokes unless the cutter is badly damaged.

Aim for consistency in top plate cutting angles and in cutter lengths. Check the former by comparing against the service mark on the top plate of the cutter, or if necessary, with a filing guide. Check the latter by measuring by eye, with a caliper, or when the cutters are of uniform length to begin with, counting the number of file strokes applied to the master cutter and use the same number on each cutter. After you've filed all the cutters on one side of the chain, along with the depth gauges/rakers (when necessary), reverse the guide bar in the vise (if using one) and sharpen the other side of the chain in the same way.

Sharpening Errors

IF THE SAW DOESN'T CUT properly after sharpening—chatters, bounces, wanders in the cut—check for any of the possible sharpening errors shown below and make any necessary corrections.

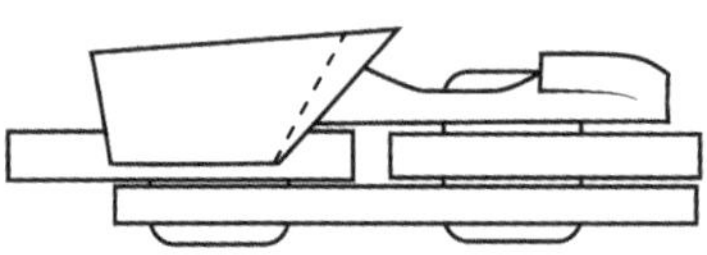

Sharpening angle too acute/narrow (compare with dashed service mark)

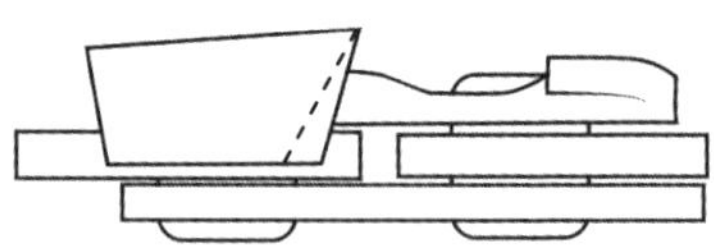

Sharpening angle too obtuse/wide (compare with dashed service mark)

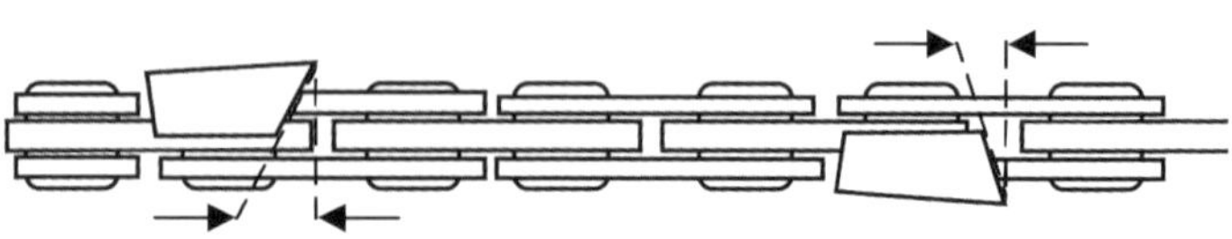

Inconsistent filing angles (angles different from one another)

Different cutter lengths (lengths different from one another)

Additional Sharpening Tips

- Sharpen the chain frequently, but not unnecessarily. "It's always better to file a little bit and often, than to file a lot, occasionally."[9]
- Use the entire length of the file during the forward stroke to extend file life and rotate it frequently to prevent uneven file wear.
- Periodically clean the file by gently tapping it or wiping it off with a rag or your pant leg. Better yet, use a file card—a metal bristled brush specifically designed to clean files.
- To sharpen a saw chain that is off the chain saw, place the chain on a spare guide bar secured in a vice.

Sharpening Procedure (Part 2): Filing the Depth Gauges (Rakers)

After all the cutters have been sharpened it's time to determine if the depth gauges need to be filed and lowered. Depth gauges, or "rakers," as they are sometimes called, determine how deep the cutters are allowed to penetrate into the wood. They do not, as many mistakenly think, "rake," or clean out sawdust from the saw kerf.

Each time the cutters get sharpened the top cutting edge of the tooth gets lower and closer in height to that of the depth gauge. When this happens, cutting efficiency is diminished, since less of the cutter is able to enter the wood. Therefore, it is critical that the depth gauges be checked routinely and filed accordingly; usually after every three or four sharpenings.

1. ***Determine the depth gauge setting***—The recommended depth gauge setting for a chain is usually listed on the chain box or packaging (stated in one-thousandths of an inch). Some chain manufacturers also have the setting stamped on the surface of the depth gauge itself, again, in one-thousandths of an inch (as in the figure at right). The two most commonly recommended depth gauge settings are .025 inches (.65 mm), and .030 inches (.75 mm). Either setting should suffice when in doubt.

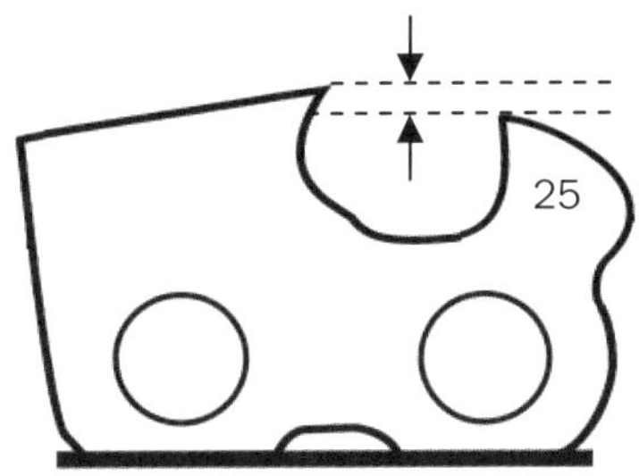

The depth gauge setting is the distance between the top of the depth gauge and the cutter's top plate leading edge.

2. ***Check and file the depth gauge***—Check the height of the depth gauge by placing the appropriate depth gauge tool on the saw chain. Press the tool against the cutter being checked with the slot positioned over the depth gauge (fig. 1). If the depth gauge protrudes above the tool it means it is too high and needs to be filed down. Use a flat file to lower the height of the depth gauges to the same level as the depth gauge tool (fig. 2). File from the inside out using strokes even with the surface of the tool. ***Note:*** Since the depth gauge tool is not made of file-hard material it should be removed before filing; use as a measuring tool only.

Figure 1

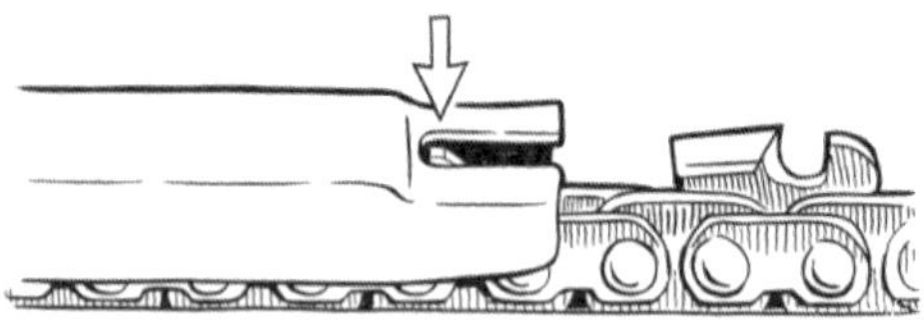

Figure 2

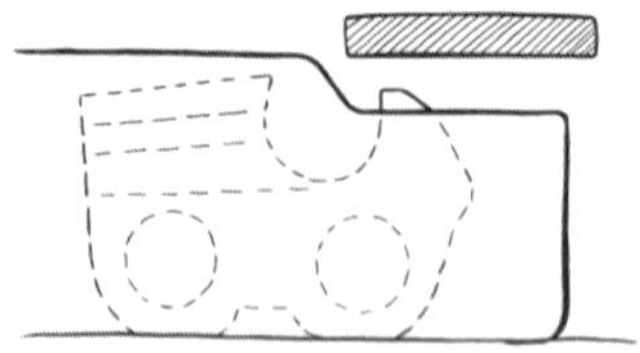

3. ***Round off the depth gauge***—After the proper height has been attained round off the front corner of the depth gauge with a few light strokes of the flat file. The goal is to retain the original shape of the depth gauge as much as possible. Be careful during filing that you do not lower the height of the depth gauge any further or accidentally touch the freshly filed cutters with the flat file.

Depth Gauge Filing Errors

IF THE SAW CHAIN PERFORMS POORLY even after proper sharpening, it should be checked for any of the following depth gauge filing errors. Depth gauges that have been left too high, filed too low, or have different settings, can cause the chain to chatter, bounce, wander in the cut, or remove little if any wood.

Depth gauge too high (file more off the depth gauge)

Depth gauge too low (file more off the cutter)

Untangling Tangled Saw Chain

UNTANGLING A TANGLED SAW CHAIN is not nearly as difficult as trying to describe, or even illustrate, how to do it—there are so many possible configurations! But knowing even a few facts and tips will help you unravel these pesky puzzles and get you and your saw back in action in no time at all. That is, of course, if you don't put the chain on backwards!

Types of tangles — Tangles always appear in saw chain in even numbers and are of two different types. They can be best described by observing the tangles in pairs, as either **matched pairs** or **mismatched pairs**.

Matched pairs are symmetrical in form, both of them appearing either inside the chain loop or outside.

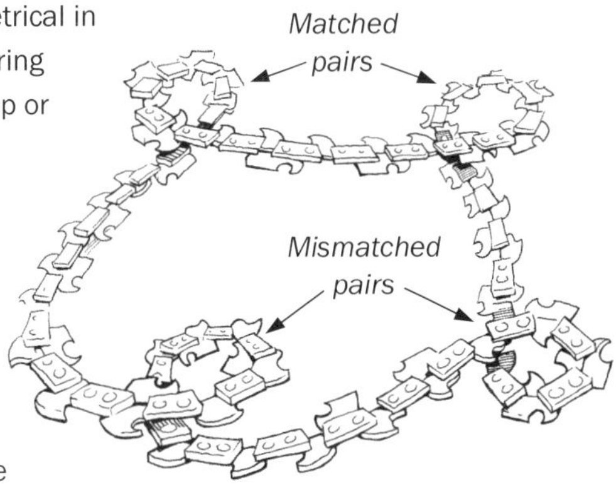

Mismatched pairs are asymmetrical in form, or dissimilar in appearance. Typically, one tangle in the pair will exist inside the chain loop and the other outside the loop—but not always.

How to untangle — There are no right or wrong ways to untangle chain if it leads to their untanglement. But if you try and untangle them in ways that are unique to their type you may have faster results.

- Matched pairs are the easiest to untangle—do these first. They are best taken out as a pair. Simply enlarge the openings of both at the same time until the two tangles merge together and cancel each other out.
- Mismatched pairs are best removed by enlarging only one of the tangles in the pair. When the opening gets large enough it will do one of two things: either invert and change the pair into a matched set, which can then be untangled per instructions above, or the tangle will simply cancel the other one out completely. The way it untangles is determined by which of the two tangles you choose to enlarge.

Note — It is helpful to lay the chain on a hard flat surface while working on it (stump, tail gate) and wise to wear gloves to avoid cutting your hands.

LOADING THE GEAR

Most of the tools needed for the day's work will already be on the work vehicle and the ones removed from the truck for servicing should be put back in their proper place. But there usually are other tools you'll also need to load that are not always carried on the truck or brought to the job. These "special" tools might include items such as chain saws with longer bars for big trees and stumps, plywood for protecting property, tires for protecting lawns, waders when working near lakes and rivers, log dollies for moving big wood, portable winches, capstan winches, or any number of other useful items.

Forgetting to bring any of these tools—the "special tools" or the ordinary-every-day tools—can be extremely costly in terms of time and money if someone has to return to the shop to retrieve them. What's more, it can pose an unnecessary risk to the job and crew if they proceed to work and make do without them.

In the same way, discovering that a tool needs repair can be equally costly and dangerous. In most cases it is the crew leader who is ultimately responsible for making sure the necessary tools are in good repair and loaded on the truck, but the ground worker should also be mindful and watchful for such things, even without being asked. Sometimes simply asking, "Is there anything else we need?" is enough to prevent a necessary tool from being left at the shop.

PREPARING WORK VEHICLES, CHIPPERS, & TRAILERS

After having been briefed on the day's work and selected, prepared, and loaded the equipment needed to perform it, it is *almost* time to head out to the job site. One last vital step of preparation remains: the pre-trip vehicle and mobile equipment inspection. To neglect any part of this inspection is to invite trouble on the road and at the job site for you, your crew, and anyone else in the area.

Many tree companies require that this inspection be performed with two people; one being the driver (who bears the greater burden of responsibility) and the other, usually a passenger. Doing so improves the effectiveness of the inspection—a second set of eyes is always more helpful—and helps to make more enjoyable what can otherwise seem like a tedious task. Using checklists will help ensure that the inspection gets done thoroughly. Without them we tend to forget even the most important things. Use the checklists that follow or adapt them to your own circumstances. Just make sure the checklists you do use are approved by your employer.

Pre-Trip Vehicle Inspection Checklist

The work vehicle is your "shop on wheels," transporting precious cargo. A thorough inspection will help ensure that it makes it to the job site safely.

Around & Under the Vehicle

- ☐ Observe the general condition of the exterior of the vehicle: body, frame, tires, wheels, undercarriage, and tool boxes.
- ☐ Check tire pressure, tread wear, and lug nuts.
- ☐ Inspect the undercarriage for any lodged debris; brake lines broken or hanging; loose exhaust components; leaks (check ground).
- ☐ Check all lights (this is when you really need an inspection partner): flashers, brake lights, running lights, low and high beam, turn signals, and dashboard lights.
- ☐ Inspect wiper blades.
- ☐ Make sure windows, windshield, and mirrors are clean (scraped of frost or ice in winter) and in good working condition.
- ☐ Check the tool box doors that they function properly and are closed and latched before leaving the shop.
- ☐ Make sure all tools stowed outside the vehicle are safely secured (ladders, pole saws, etc.).

> "Captain De Long scrutinized his weather-beaten ship...going over every valve and fitting, every strake of her long hull. He wondered where her weaknesses lurked. Were there rotten timbers? Leaky seams? The smallest flaw could mean his death, and the deaths of the men who would serve with him."
>
> —Hampton Sides
> *(In the Kingdom of Ice)*

Under the Hood

- ☐ Look for anything out of the ordinary: debris, signs of rodent activity (acorns, nests), leaks, bad smells.
- ☐ Check all fluid levels: engine oil, brake fluid, radiator coolant, power steering fluid, and windshield washing fluid.
- ☐ Inspect belts, hoses, and electrical wires for wear, cracks, or fraying.
- ☐ Battery cables for wear, corrosion, and looseness.

In the Cab

- ☐ Clean cab seats and dashboard of any trash or unnecessary debris.
- ☐ Stow safely or remove any objects that could shift while en route.
- ☐ Check that mirrors are clean and in good working order.
- ☐ Clean the windows and windshield if necessary.

Stowed Onboard: Safety & Service Equipment

- ☐ Water cooler (clean, filled, and secured)
- ☐ First-aid kit (stocked, accessible, and its location known to all the crew)
- ☐ Wasp/hornet spray
- ☐ Jack and other tire changing tools
- ☐ Traffic signs, flags, safety cones
- ☐ Four-way flashers
- ☐ Flagging tape
- ☐ Wheel chocks
- ☐ Fire extinguisher
- ☐ Maintenance tools
- ☐ Extra fuses
- ☐ Flashlight
- ☐ Maps
- ☐ Phone charger
- ☐ Tire chains (when applicable)

Pre-Trip Chipper & Trailer Inspection Checklist

Use the checklist below to inspect chippers, trailers, and other mobile equipment to be transported and used at the job site. In most cases the chipper or trailer should be hooked up *before* performing these inspections (to test lights for example). See pages 52-53 for hook up procedures.

Chipper & Trailer Inspection

- ☐ Observe general condition of the chipper or trailer. Look for any broken, missing, or loose parts, or lodged debris in the frame.
- ☐ Check tire pressure, tread wear, lug nuts (missing? loose?).
- ☐ Check all lights to make sure they are working properly.
- ☐ Check that the jack stand is in good working order and stowed in place.

For chippers only:

- ☐ Make sure you know where the operator's manual is located.
- ☐ Check crankcase oil and engine coolant levels.
- ☐ Check hydraulic fluid level and all hydraulic components for leaks or damage.
- ☐ Check fuel tank.
- ☐ Check upper feed control bar for proper operation.
- ☐ Check lower feed stop bar (if available) for proper function.
- ☐ Inspect shields and guards.
- ☐ Check machine for proper lubrication (grease).
- ☐ Look for any missing, loose, or broken parts.
- ☐ Check drive belts for proper tension.
- ☐ Check condition of all safety and operating decals.

Note: See pages 157-160 for chipper setup and operation.

For trailers only:

- ☐ Make sure the load/equipment being hauled is secured properly and all tie-downs (chains, rope, or straps) are in good condition.

Hooking Up Chippers & Trailers

Failure to correctly hook up chippers and trailers can lead to serious damage to the towing vehicle, the unit being towed, and to other vehicles on the road, should the unit come unhitched while driving. Take your time and get it right, and even then, double check your (or another's) work.

Backing Up to a Chipper or Trailer

Being the driver's extra pair of eyes.

- ☐ Always have a spotter help the driver back up to the trailer or chipper.
- ☐ Learn the necessary hand signals to guide the driver to the hitch of the unit being towed: left, right, straight back, clear, keep coming, distance to hitch, and stop.
- ☐ The spotter should always be positioned out of harms way and visible to the driver, preferably in the driver's side mirror. The driver should be able to see the spotter at all times.
- ☐ Once the hitch is positioned directly over the ball, the driver should put the vehicle into park while keeping a foot on the brake the entire time. Now the spotter may hook up the trailer or chipper safely.

Hooking Up Chippers & Trailers

- ☐ Lower the trailer hitch onto the vehicle hitch ball until the full weight of the trailer is on the ball.
- ☐ Close the coupler clamp and lock the coupler on the ball by putting the hitch pin into the hitch coupler lock.
- ☐ Retract the jack, securing it into transport position or removing it entirely and stowing it in a suitable place.
- ☐ Hook up the tow chains. They should be crossed under the tongue and twisted to shorten or lengthen such that the chains are not dragging on the ground or bind while making turns.

- ☐ Plug in the electrical connections (lights and trailer brakes).
- ☐ Connect the emergency brake cable. Make sure there is enough slack in the attachment that sharp turns won't activate the brake.

For chippers only:

- ☐ The chipper should be level during transport—adjust the hitch height on the chipper and/or transport vehicle accordingly.
- ☐ Position and lock discharge chute in the forward position.
- ☐ Fold up feed table and secure latch to transport.

BEFORE LEAVING THE SHOP

Just one last list of preparations to perform before leaving the shop. These are mostly, but not entirely, performed by the driver of the work vehicle.

- ☐ Start the truck, warm the engine, and look, listen, feel, and smell for any potential problems—smoke, leaks, burning smells, unusual vibrations, knocking or rattling sounds—anything out of the ordinary. If there are any issues turn off the engine and inform your supervisor.
- ☐ If you haven't already done so, stow safely or remove any objects that could shift while en route.
- ☐ Employ the A-B-Cs—***Adjust*** seat and mirrors, ***buckle*** your seatbelt, and ***check*** that your passenger is also buckled up and ready to go.
- ☐ Check gauges that they are working and in their safe operation range.
- ☐ Turn headlights on. Don't assume they'll come on automatically.
- ☐ Turn the radio off or set volume low enough that you can converse with the other passenger(s). This is a good time to test the horn as well.
- ☐ Plan your route to the job site (if you haven't done so already) *before* leaving the shop. Looking at maps and apps while driving means taking your eyes off the road and is therefore, an unacceptable activity. While en route, the passenger can take on the navigational responsibilities.

With inspections completed and gear and crew loaded you're now ready to drive to the jobsite. As you leave the shop your role as a ground worker will noticeably shift from that of preparation and inspection to that of committed watchfulness on the road—the topic of the next chapter.

The Ground Worker's "One Thing"

THE FIRST TIME I heard about the "One Thing" concept was while watching the hit comedy *City Slickers*. It was described in a conversation between the trail boss, Curly (Jack Palance) and the city slicker, Mitch (Billy Crystal) as follows:

Curly: Do you know what the secret of life is?

Mitch: No. What?

Curly: This. (He holds up one finger).

Mitch: Your finger?

Curly: One thing. Just one thing. You stick to that and everything else don't mean sh*t.

Mitch: That's great, but what's the "one thing"?

Curly: That's what you've got to figure out.

If I could offer only one piece of helpful advice to ground workers—to all tree workers for that matter—which would serve as their "One Thing" for success (and survival) in tree work it would be this: **PAY ATTENTION!**

To pay attention is to be mindfully aware of what's going on around you—at the shop, on the road, and at the job site—at all times; it means being on the lookout for hazards and obstacles to avoid, for ways to be useful, for things to learn, and then to act accordingly.

When you make "*pay attention*" your every day, moment-by-moment, "One Thing" motto and practice, you'll discover how it affects everything you do as a ground worker. The more alert you are for potential hazards, the more likely you are to notice and prepare for them, which makes work (and driving) safer; when you are watchful for ways to be useful, you become more productive, and as a result, a more valuable team player; and when you are attentive for opportunities to learn, you gain more knowledge and skill, and thus become a more competent and effective tree worker.

Paying attention, however, does not come naturally to most of us. Studies have shown—along with personal experience—that our minds are easily distracted and prone to wander. Therefore, learning the art of paying attention will require the same disciplined practice that it takes to master any other skill in tree work. Turn to pages 74, 100, and 118 to learn more.

3

The Role of the Ground Worker...

On the Road

Driving to the job site is as much a part of the job as the work you perform once you get there. It can also be just as dangerous. The types of hazards on the road differ only in kind, not in number. The hazards are endless: bad road conditions, bad weather, bikers, bicyclists, pedestrians, stray animals, unsafe drivers, and road distractions ranging from billboards to sunbathers. These are only a handful of hindrances to reaching your destination safely. With so many hazards on the road, it's a miracle that tree workers reach their destination at all.

Aside from relying on a miracle—or luck—there are many things drivers and even passengers can do that contribute to a safe trip to and from the job site. In this chapter you'll learn what those things are. Going through "drivers' training" again probably doesn't sound as interesting as all the other cool things you've learned in this book thus far (and will be learning in chapter 4). But think about the consequences if you were the cause of an accident that led to your own death or injuries, or led to the death or injuries of someone else.

It is when we realize the potential consequences of our driving mistakes that we take seriously our responsibility as a driver, or even as a passenger. As in all areas of life, choices have consequences. Good and bad. Read on to find out how to make the best choices that will get you there, and back again—alive, uninjured, and without incident.

THE DEFENSIVE DRIVER

The goal of every driver should be to follow local traffic laws and practice proper driving technique. But even if you do everything right, you can still be in an accident because of the actions of others or things you have no control over—adverse weather, traffic, and road conditions, just to name a few. The only thing you have control over are your own choices and actions. A defensive driver then, is someone who makes the best choices in light of the present circumstances. A defensive driver will be aware of potential hazards and take the correct action to deal with them before they become a problem. Becoming a good defensive driver begins with making good driving practices a matter of habit. What follows are ten of the most important driving habits you can develop.[10]

Pay Attention! The Habit that Rules Them All

As you just learned from page 54, *paying attention* is the most important habit of tree work you can develop (according to the author anyway). This is most certainly true when it comes to driving to the job site. A good defensive driver needs to maintain a laser-focus on the road and surrounding area, as well as his or her own actions while driving—lives depend on it! Being watchful on the road may not guarantee an incident-free ride, but it will drastically lower your odds of having one.

Habit #1: Wear a Seat Belt

Hopefully, the cautionary advice to "buckle up for safety" has not become so familiar that you have lost sight of its importance. It is a statistical fact: wearing seat belts saves lives. They increase your chance of surviving a collision by about 50 percent. That's because they keep you in the vehicle, rather than flying through the windshield, dead on the road. Keep that image in mind to help you remember to always buckle up. Also, make it a habit to remind your passenger to buckle up as well—even if it's the boss.

Habit #2a: Recognize Hazards

The most important reason for paying attention while driving is to recognize potential hazards so you can apply the correct preventive action. To do this most effectively you will need to utilize not just your eyes but also your sense of hearing, smell, and even touch or feeling.

Sight—By using a visual technique called scanning, you can take in the big picture of road activity—not just what's going on ahead of you, but *around* you as well. As you scan the area ahead, look to the next intersection, block, overpass, hill, or curve; avoid focusing on any one thing too long. Look at the sidewalks and front yards; finally, look behind you, checking your mirrors frequently.

> "You see, but you do not observe"
>
> – Sherlock Holmes
> *(A Scandal in Bohemia)*

Hearing—There are many potential hazards which may be detected only through hearing: the approach of an emergency vehicle, a car honking a warning, a strange sound coming from the engine. If you can't hear them because of loud music playing, among other things, you won't know of the potential hazard. If you must listen to the radio, keep the volume at a level that allows you to hear what's going on around you. (I once had a young employee who, for lack of instruction on my part, had the radio so loud that he never heard, or felt, the chipper pop off the hitch, drag on the chains 100 feet before unfastening completely and then fly into the ditch. Only after arriving at the shop did he notice something was missing.)

Smell—The smells you should be most concerned about and inspect are the ones coming from your own work vehicle: the smell of burned rubber (engine hoses, belts), burned carpet (brakes), syrup or sweet smell (leaking coolant), sulfur or rotten eggs (catalytic converter), hot oil, or gasoline.

Touch—Even the sense of touch or feel is vital to attentive driving. If, for instance, the road is icy, you should feel the vehicle slide slightly; if tires are out of alignment (or going flat) you should feel the vibration.

Habit #2b: Act Correctly

Whatever the detected hazard may be, you must apply the correct preventive action. If you see a Road Work Ahead sign, slow down and be watchful for workers. If you hear an emergency vehicle siren, check your mirrors, signal your intentions, and pull over to the side of the road. If it's raining, turn your wipers on, slow down, and put plenty of distance between you and the vehicle ahead. If you smell gas, pull over as soon as possible and check for leaks. It's not enough, however, to make the correct action, you need to apply it *in time*—before it's too late.

Use the "What If?" Strategy

ONE WAY to improve your observation and reaction skills while driving is to apply the "what if?" strategy. It begins by imagining a possible driving or traffic scenario you could encounter (the *what if* part of the equation) then answering the "what would I do if it did?" question.

For instance, "what if" the load of lumber on the trailer ahead of you were to suddenly come loose and scatter boards all over the road? What would you do? Or better yet, how could you prepare in advance to avoid the situation entirely? Perhaps you should pass the vehicle at the first chance you get. Or, if the vehicle is already traveling the speed limit, maybe you should slow down to put more distance between your vehicle and the other one. At the very least, plan your escape route should it happen. After thinking through that "what if?" scenario, do another. Applying the "what if" strategy" will train your mind and eyes to detect potential danger and prepare you to act correctly were it to happen. It will also help keep you awake in the process.

Habit #3: Avoid Distractions

It's quite simple: you can't pay attention to the road if you're doing something else while driving. Fumbling with food, fiddling with radio dials, and talking and texting on phones, are the most common driving distractions. Taking your eyes off the road to do any of these things, even for a moment, can cause your vehicle to drift into oncoming traffic and result in a head-on collision—the most deadly of all accidents.

Take cell phones for example. It takes 4.6 seconds to text an average text message—dialing a 10-digit phone number takes a similar length of time. If you want to know how much can happen in that short window of time, try this the next time you are a passenger in a vehicle:

Close your eyes for a count of five seconds and imagine all the bad things that could have happened if you were driving. Way too many. If you think hands-free phone devices are the solution, think again. Studies have shown that these too can cause mental distractions and therefore, offer no safety benefit when driving. If you need to text, answer, or make a call on your cell phone, either have the passenger do it or find a safe place to pull over and take care of business there.

The same goes for food and drink: enjoy your food *after* arriving at your destination so you can focus your attention on getting there. As for listening to music, keep the volume at a reasonable level so you can hear the passenger give warnings or instruction and any possible road hazards. One possible solution for keeping your hands off dials is to use a playlist. And speaking of dials, learn what all the buttons and controls on the dash are for *before* you head down the road.

Habit #4: Plan Your Route

Another common driving distraction occurs while looking at maps, or more likely, your cell phone to find your way to the job site. Here again, your eyes and mind are diverted from what's happening on the road. The best practice here is to plan your route ahead of time—at the shop or the last job you were at—and have the passenger give directions. If driving alone, use a navigation app with voice-directed guidance or pull over in a safe place to find your way.

> "When you come to a fork in the road, take it."
>
> –Yogi Berra

Habit #5: Avoid Driver Fatigue

Driving while fatigued can be as risky as drunk driving. Yet many of us do it anyway. Although there are all sorts of alertness tricks drivers use—drinking coffee, turning up the radio, opening the window, slapping yourself—none of them are real cures for drowsiness. If anything they can give you a false sense of security. Getting adequate sleep—seven to nine hours a night—is the only sure way to overcome driver fatigue. Eating healthy foods and staying hydrated are also important factors in avoiding fatigue. Avoid medications that cause drowsiness, such as some cold and allergy medications. And do we even need to mention avoiding alcohol and drugs?

Even after taking these precautions, fatigue and drowsiness can set in—especially after a hard day of tree work. What then? First, recognize the signs of drowsiness: frequent yawning, heavy eyes, blurred vision, swerving toward the shoulder or over the center line. If you experience any of these symptoms ask your passenger to take over driving. If you're alone, and find none of the alertness tricks to be of help, pull over somewhere and take a brisk walk. Fresh air and light exercise work wonders. If necessary, take a short 10 or 15-minute nap. Maybe let someone know first.

Habit #6: Control Your Emotions

If ever there was a time to exercise patience and self-control it would be when driving. There is no end to the number of driving situations that can trigger anger. And angry drivers are dangerous drivers. They are much more prone to road rage and aggressive driving behaviors like speeding, tailgating, erratic lane changing, and shoulder passing, all of which put yourself and others in danger.

So what can you do to avoid losing your cool while driving? Focus on what *you can* control, not on what you can't. You can start by controlling your own emotions when things don't go well. Self-control looks like this: when another driver honks at you for driving too slow and flips you the middle finger as they pass, don't return the rudeness. Don't even make eye-contact. Expect such things to happen. Accept it. Laugh even. Try making their behavior less personal by imagining some difficulty the other driver is going through (put a name to it) and that you just happen to be the anonymous person to receive their anger (see also, pages 65-66).

Habit #7: Don't Speed

The number one unsafe driving behavior that contributes to collisions and violations is improper speed. In recent years, speeding has been the major contributing factor in over twenty five percent of collision deaths. Speeding is not just driving over the posted speed limit, but also driving too fast for the existing road and traffic conditions. People have all sorts of excuses for speeding: "I was late for work. I didn't know the speed limit. Everyone else was speeding. I do it all the time and nothing happens." But things *do* happen when you speed: tickets, fines, collisions, damage to property, lawsuits, injury or death to you, your crew, or others. Here's the bottom line: speeding is a choice. Choose to slow down.

Habit #8: Maintain Proper Following Distance

Following too closely behind the vehicle ahead of you creates two hazards. First, it limits your view of the road ahead, making it difficult to see and anticipate any potential problems or hazards. Second, tailgating increases the risk of a rear-end collision should the vehicle ahead of you apply their brakes. What's more, tailgating is considered a form of aggressive driving and a ticketable offense. To avoid a rear-end collision, or getting a ticket, apply the ***four-second rule*** to maintain a safe following distance as follows:

1. Watch for the vehicle ahead of you to pass a fixed object, such as a sign, mailbox, or tree.
2. As the vehicle's rear bumper passes the object, begin counting: 1001, 1002, 1003, 1004.
3. Your vehicle should not pass the same fixed object before you finish counting to four.

The four-second rule should be increased by at least one-second for each hazard that may be present. If for example, you are hauling a heavy load along a winding road, increase your following distance by two seconds (one for each hazard) for a total of ***six*** seconds. Also increase your following distance when driving in adverse weather conditions, a construction zone, or even for things such as driving into the sun.

Habit #9: Signal Your Intentions Early

When making a turn or a lane change, it's important to signal your intentions early so others have time to respond safely. If you don't, you could get a ticket and fine, but more importantly, be the cause of an accident. When making a lane change, signal your intentions with six or more clicks of the turn signal in advance. Check your mirrors and blind spots before changing lanes. When making a turn, first get into the proper lane—again, making the lane change as described above. Once you're in the proper lane, signal your intentions early—100 feet is a good minimum distance—checking your mirrors and blind spots first.

Habit #10: Exercise Caution at Intersections

More than half of all collisions occur at intersections. Some of the reasons why intersections are so dangerous is because (1) they are visually complex—there's a lot going on there, (2) there are multiple hazards that need to be evaluated, (3) traffic patterns are continually changing, and (4) they impose time pressure for making decisions. Reason enough to approach them with the utmost of caution. When approaching intersections, apply the cautionary practices that follow:

1. ***Approach at a safe speed,*** ready to stop for such things as a changing traffic light, a turning or braking vehicle, or a pedestrian in a crosswalk. ***Never*** change lanes or pass someone at an intersection.
2. ***Stop at a safe distance*** from the vehicle ahead of you to allow for an escape path if needed. A safe stopping distance is at the point where you can see the vehicle's back tires where they touch the pavement.
3. ***Don't jump the gun*** when the light turns green. If you are the first one at the intersection and the light turns green, before accelerating look left, look right, look straight ahead, and then left again. Keep your foot on the brake as you do so. If another vehicle is ahead of you, watch for its tires to move, then count, 1001, 1002, before accelerating. In both cases this will delay your acceleration by two seconds, time enough to ensure that the intersection is clear to proceed.

OTHER DRIVING CONSIDERATIONS

Backing. Backing vehicles is not only challenging—especially with an attached chipper or trailer—but is also one of the leading causes of collisions in the tree care industry. If you must back up, have a coworker positioned behind the work vehicle (also known as a *spotter*) help guide and direct you (see page 52). If you are alone, first perform a "circle check" before backing, which means get out and walk around the vehicle and look for potential hazards. Check under and above the vehicle as well.

Parking. Here are six important rules for parking the work truck: (1) don't park in front of a fire hydrant or too close to an intersection, (2) don't block an entrance or driveway, (3) leave the vehicle in the lowest gear, or in park, and chock the wheels, (4) when parking facing uphill, turn the wheels away from the curb and when parking facing downhill, turn the wheels into the curb, (5) lock the cab and toolboxes when leaving your vehicle and remember to take the keys with you, (6) when applicable, set out safety cones per instructions on page 68.

Driving in snow or ice. Before driving, remove all snow and ice from vehicle windows and windshields and make sure the wiper blades are functioning properly. Once you're on the road, drive slowly and increase your following distance to six to ten seconds; accelerate and decelerate slowly; do not use cruise control; and try to avoid stopping while going up a hill.

If You Have an Accident

WHETHER YOU EXPERIENCE a collision as a result of your own mistakes or that of another driver, what matters now is what you are going to do about it. Rather than blaming yourself or shifting the blame, shift into problem solving mode and apply the procedure below:

1. Stay calm!
2. Stop immediately, or as soon as practical and safe.
3. Make sure that all people involved in the incident are removed from danger, and take precautions to prevent further collisions.
4. Determine if anyone involved has been injured, and summon the appropriate authorities.
5. Exchange license, registration, and insurance information with the other party in the accident.
6. Notify your employer of the collision.
7. Follow your company's policy about making verbal statements concerning the accident.
8. If your vehicle strikes an unattended vehicle and you are unable to locate the vehicle's operator, leave written information in a secure and conspicuous place on the unattended vehicle.
9. Obtain the names and phone numbers of witnesses if necessary.
10. Follow your company's policy for filling out accident report forms, etc.

** This procedure has been borrowed from Ground Operations Specialist, part of the Tree Care Academy series of training programs from TCIA.*

THE HELPFUL PASSENGER

The phrase "riding shotgun" was once used to describe the guard who rode next to the driver of a stagecoach. With shotgun in hand, his primary job was to ward off hostile bandits. Reaching their destination safely depended much on the guard's watchful eye and skillful accuracy with a gun. In a similar way (minus the gun of course), the passenger of the work vehicle plays a vital role in helping the driver reach the job site safely. Riding shotgun in the work vehicle is not a time to nap or obsess over your phone, but to help the driver in the following four ways:

Help the driver observe hazards—The driver of the work truck has a lot to look for and think about. You can help by being the driver's extra eyes, ears, and even nose (see page 57). Notice especially road and traffic activity that's on your side of the road. Periodically check the mirror's blind spots. When you approach a stop sign, give the driver a verbal signal, such as "all clear my way", when the coast is clear, or a "car coming" when it's not. Also, be alert to strange sounds or smells that a road-focused driver may not notice.

Help the driver navigate—Never assume the driver knows where he or she is going. Help the driver navigate by reading the map and giving clear directions (turn the music off) well in advance of any needed lane changes and turns. Check the work order for instructions for where you should park or approach the job site once you arrive.

Help the driver concentrate—In other words, *don't* distract the driver with loud music or phone videos, by talking constantly (unless you're really interesting), or by rummaging around in the glove box or your lunch box. Don't give the driver any reason to look at "what you're up to."

Help the driver stay awake—Keep an eye on the driver's eyes. After a long day of tree work, collapsing into the comfortable seat of the work truck is often a prelude to heavy eyelids. If it appears that nap time is not far off, offer (insist?) to take over driving. If that's not an option, engage the driver in conversation. Ask open-ended questions—What did you find most challenging about the job we did today?—rather than closed-ended questions you can answer with a yes or no. Consider the taboo topics—religion and politics—that will most certainly revive the driver awake.

What to Do When People Are Rude

ONE OF THE RECURRING THEMES in this book is building character through difficulty. One of the greatest of these is having to deal with rude people. Which is to say, people who are uncaring, unkind, and just general jerks. Rude people are everywhere. You work with them, share the road with them—you might even be one yourself! How you deal with their rudeness will determine if the situation continues, worsens, or in the end, works out for good. Here are some tips that will lead to the latter of these.

Expect rudeness — One of the secrets of contentment is to expect things *not* to go as you would like them to. As it relates to this topic, expecting rudeness to occur at anytime helps make it more bearable when it happens. The surprise factor is eliminated. If, on the other hand, you expect traffic to kindly yield to your welcome presence as you come down the entrance ramp, or that your work place should be a haven of kindness and courtesy (not that it can't be), you will most certainly become offended, frustrated, and angry.

Overlook rudeness — Recent studies, along with simple observation, have shown that people today, particularly young people, are becoming increasingly more fragile. They have become hypersensitive to the things people say and do that are not to their liking. Unfortunately, this fragility makes them—along with those who have to listen to them—more unhappy as a result. Incidentally, it has also been shown that workers who are easily offended make terrible employees—they're so often distracted by their feelings about some offense that they can't effectively focus on their work.

A better option to being easily offended and becoming angry and hurt is to consider overlooking the offense. Indeed, it is a very counter-cultural response, but one that reflects humility and maturity. When you choose to overlook an offense, possibly even forgiving the offender as one who doesn't know any better, you can chalk it up as a personal victory of strength—not weakness. However, if the rudeness was a verbal one, you might also consider that just maybe the hurtful thing said was true, only poorly presented.

Confront rudeness — Let's begin by stating what type of people you should not confront. One: another driver. There is simply no way to effectively deal with rude drivers other than overlooking the offense. Two: customers. If you have a problem with a customer either ignore the offense or bring it to your crew leader's or boss's attention. *(Continued on the next page)*

Even confronting your boss can be dangerous if handled poorly, but that's one for another book. So at this point, we're left with confrontations with coworkers. For a productive outcome consider these suggestions:

1. ***What to do.*** First off, confrontation does not mean getting even with the person who has offended you—lashing out in anger, returning rudeness with rudeness, hurt for hurt. Productive confrontation is about resolving a problem calmly, thoughtfully, politely, but yet, assertively. Start by taking a few deep breathes and counting to ten, or whatever it takes to cool off and calm down. Only then can you think clearly about what to do next. At this point you might decide to postpone confrontation until a better time or place. You might even ask the other person when a good time would be to talk about the matter. This will also give you time to better prepare what you want to say. But sometimes, the best time to confront is now.

2. ***What to say.*** Before you say anything, *think* first. Make *thinking before speaking* one of your mottos in all of life. Doing so will be one of your best investments of time, especially when facing difficult people. Why is that? Because once words pass through the door of your lips, they're out there to heal or hurt, help or harm. There's no do-overs. Also, pay attention to your tone of voice—people always respond to your tone of voice.

   Now, to be a bit more specific about what to say, consider the S.I.R. sequence: **S**—state the facts; bring your complaint to the offender's attention; **I**—inform the other person the impact that the offense had on you. As much as possible, avoid leading with a "you" statement—it sounds like an attack. Say "I was offended when you" rather than "you offended me when."; **R**—request that the rude behavior not be repeated.[11]

3. ***What to expect.*** If you've chosen your words wisely and stated your case politely (but firmly) and respectfully, you might receive the same in return. Both are contagious. You might even get an apology. But don't count on either. It's more likely that the other person will shift the burden of blame back to you. And that's OK. At least you no longer have to carry the burden of unresolved anger and you let the other person know you're not one to be trifled with easily. The offender might think twice next time.

** In cases of extreme workplace rudeness (bullying), it's best to take it to the top—to your boss or crew leader. At the very least tell a wise and trusted coworker or two, "for the record," and to get their opinion on the matter.*

4

The Role of the Ground Worker...

At the Job Site

Nowhere is the ground worker more needed, and thus more appreciated, than at the job site. Sure, their contributions at the shop and on the road are important, but it's here at the job site where ground workers are undeniably indispensible. Which is to say, that though ground guys matter all the time, they especially matter at the job site. From controlling traffic to talking with customers, from roping down limbs to felling trees, and from moving limbs and logs to the wood chipper to raking the lawn, the tasks the ground worker performs is seemingly endless.

This is why this chapter is the longest. The material that follows is grouped into two parts: *job planning and preparation* and *job execution*. Or if you prefer, "planning the work and working the plan," as the popular work axiom goes. Either way, the success of the job hinges on the skillful execution of both.

PART 1: JOB PLANNING & PREPARATION

As you've already learned from the last three chapters, work planning and preparation begin at home, continue at the shop, and even on the road, as various details of the work order are discussed amongst coworkers. Once you have arrived at the job site, plans and preparations continue with the same goal as when you first left home—to have a safe and successful day of tree work. That is, completing a job to the specifications of the work order, the details of the work plan (more on that later), and to the satisfaction of the customer and your employer.

UPON ARRIVAL

When you first show up at the job site, try to imagine how peaceful and quiet it probably was before you and your crew showed up, especially if it was in the morning hours. As much as you can, try and preserve that calm, even though it probably won't last long. Also, upon your arrival, you must take every measure possible to ensure the safety of everyone you will encounter—yourself, your crew, the customer, pedestrians, and vehicular traffic. This begins the moment you arrive at the job site.

Put Your Helmet On!

As soon as you get out of the work truck, find your helmet and put it on. Along with protecting your head from any number of unknown hazards, it sets a good example for your coworkers, and sends a clear message to the customer and public that you are authorized to be on the property and not some unwanted ne'er-do-well, thief, or salesman. So be professional: put your helmet on!

Controlling Traffic

If you will be parking work vehicles and equipment on a road with pedestrian and vehicular traffic, it is essential that you make the scene safe. This will involve such things as setting up warning signs, traffic cones, and barricades in suitable locations.

Placing Traffic Cones

- *Always face traffic when placing and removing cones.*
- *Always use the vehicle's four-way flashers, and if equipped, beacon lights when setting up cones and signs.*
- *Always chock trucks and detached chippers with at least two wheel chocks.*
- *Space cones on roadways the same number of feet apart as the speed limit (or closer).*

However, don't trust in these alone to stop other drivers—people are frequently distracted and inattentive and will drive right around or through your attempts to stop or warn them. Therefore, it is often necessary to have flaggers posted to keep traffic flowing smoothly or intercept any vehicles from entering the work area.

As a tree worker who works near roadways, you also need to keep yourself safe. You can do so by wearing high-visibility apparel. This will significantly reduce your chances of being struck by vehicular traffic. Fluorescent yellow-green and orange-red are the most effective colors for this purpose. But even while wearing bright clothing, never assume that vehicular traffic will see you; rather, always pay attention to them.

Working With the End in Mind

WHEN A WORK CREW knows the goals of the job *before* they begin working, they can more effectively take the proper steps to achieve them. That's what it means to work with the end in mind. So what are the end goals to which all tree workers work? At the very least it is to complete a job in such a way that:

1. No one gets injured or killed.
2. No property gets damaged or destroyed.
3. No equipment gets damaged, destroyed, lost, or forgotten at the job site.
4. The work is completed according to plan and in the allotted time frame specified on the work order.
5. The work team functions together in a skillful, coordinated, and agreeable fashion—like a well-oiled machine.
6. The customer and your boss are satisfied with the crew's performance, and, as much as possible, positively influenced by their presence.

The success of a job can be measured against how many of these goals were achieved. The ideal of course, is for all six of these goals to be flawlessly achieved. But, as we all know, stuff happens. Things go wrong. People screw up and make mistakes, preventing even the best-laid plans from materializing into the desired results. But, with a clearly defined *end in mind*, job success, as described above, is more likely to occur more frequently.

Scouting Out the Driveway

If vehicles and equipment are to be parked and operated on the customer's property, but off from the main roadway, it's wise to scout out the driveway first. The main objective here is to determine if there is enough room on the property to maneuver vehicles and equipment into work position and to safely leave the job site afterward.

If there isn't sufficient turn-around room, you'll need to determine if it is more efficient to back the work vehicle into the property at the start of the job rather than later. A thoughtful job estimator would have already considered this factor and made a note on the work order. In either case, always be on the lookout for overhead wires, eaves of structures, and any other hard-to-see hazards, and *always* have a spotter when backing up a vehicle.

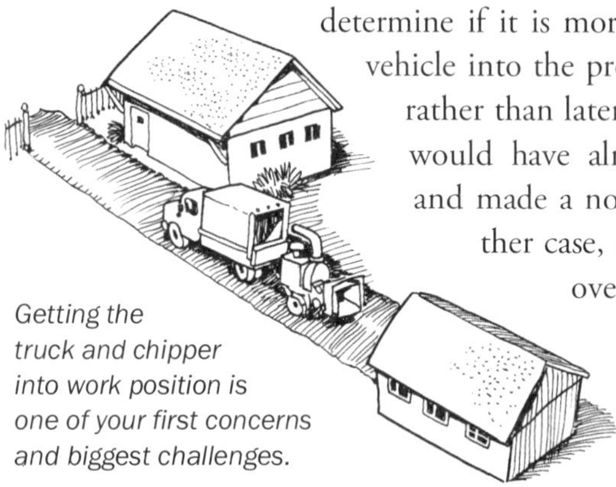

Getting the truck and chipper into work position is one of your first concerns and biggest challenges.

Checking in With the Customer

It is both a safe and courteous practice to inform the homeowner, and in some cases even the neighbors, of your presence at the job site. No customer should have to be made aware of your arrival by the scream of a chain saw and wood chipper. That said, if it is early in the morning, do as much prep work as possible before starting noisy equipment. Many tree companies make it a practice to contact the customer before the work crew even arrives. This not only eliminates the surprise factor, but gives customers the opportunity to move their vehicles out of the way.

Even if the customer has been notified beforehand, it remains important that they be informed of your arrival and your work plans. They should be instructed to keep small children and pets indoors if possible. If they do desire to watch the work, they should be shown where they can safely do so following the guidelines on page 83. Remember, as a representative of your company, you only have one chance to make a good first impression on your customer—be courteous, friendly, and respectful.

INSPECTING THE JOB SITE

After performing the preceding precautions, the next step is to inspect the job site for hazards and obstacles. A work hazard is anything present at the job site that may pose a risk to people or property. An obstacle is any object you wish to avoid damaging, or which hinders your ability to work on a tree. The presence of either greatly influences how the work plan is prepared and the work is performed, as well as how hazards and obstacles will be protected or avoided. Many site hazards and obstacles will have already been identified by the job estimator and indicated on the work order. Invariably, though, some are missed and others turn up. Even though the crew leader is primarily responsible for overseeing the site inspection, it's essential that all tree workers know how to perform one.

There are at least six different factors to address during a site inspection: (1) ground hazards, (2) tree hazards, (3) electrical hazards, (4) people and pets, (5) weather, and (6) obstacles. Having a sharp eye is crucial to finding any of these potential hazards. Sometimes, however, even obvious hazards go undetected when the inspection becomes routine. That's why it's always helpful to ask questions during inspection. For instance, asking yourself, or a coworker, "what could kill us today on this job?" or, "what objects do we not want to spend our next paycheck replacing or repairing?" can wake you up to dangers that might otherwise go unnoticed.

What objects do you not want to spend your next paycheck replacing or repairing?

Ground Hazards

Start your inspection by looking for any objects or terrain features that could be a threat to workers or interfere with work operations. Clear whatever hazards that you can before starting work. Look for any of the following:

- ☐ Poisonous plants
- ☐ Ground nests of stinging insects
- ☐ Wet and muddy areas
- ☐ Dense undergrowth in the work area
- ☐ "Slip, trip, and fall" hazards: steps; ground debris (including pet and animal excrement), stumps, logs, and rocks; uneven, steep, or rocky ground; slippery ground caused from rain, ice, or snow

Tree Hazards

The trees you will be most concerned about identifying as hazards are the ones on which you'll be working—those specified on the work order. But be alert to hazards in other trees as well. If they are close enough to the work area, they too will have to be dealt with. It's also possible that these hazards were overlooked by the job estimator and could be serious enough to warrant pruning or removal. Inform your crew boss if this is the case.

Begin your assessment from a distance, as you first approach it. Many hazards, such as widow-makers (a lodged branch or treetop), or an overhead electrical line, may be hidden when viewed from directly below the tree. Once you've reached the tree, continue to inspect all sides of the tree, starting with its base, along the trunk, and into the canopy. Be watchful for:

- ☐ Limbs in the canopy that are dead, weakly attached, or completely broken off and hanging in the canopy (widow-makers).
- ☐ Cracks, cankers, and cavities which could indicate tree decay and potential tree failure.
- ☐ Signs of trunk decay, such as mushrooms, conks, loose bark, and insect activity.
- ☐ Signs of tree failing or falling, such as tree leaning heavily and soil mounding on one side.
- ☐ Bees, wasps, and hornets which frequently build exposed paper nests in the tree canopy or nests hidden in trunk cavities.
- ☐ Animals, such as birds, bats, squirrels, raccoons, etc., which frequently dwell in hollows in trees, and bring attention to the defect as well.

Neglecting to spot any of the above tree and site hazards could be disastrous.

- ☐ Vines, which can be poisonous and also hide other hazards.
- ☐ Foreign objects in the trunk, such as nails, wire, or concrete, that could interfere with felling and bucking cuts.

Electrical Hazards

ANSI Z133-2017 (4.1.2) states that "all overhead and underground electrical conductors, guy lines, pole grounds, and communication wires and cables shall be considered energized with potentially fatal voltages." Contact made (direct or indirect) with any type of these energized conductors or electrical fixtures is one of the leading causes of death and injury in the tree care industry. Avoidance is key—both in preventing personal injury or death and in avoiding costly repairs if a conductor is damaged. Avoidance begins by locating electrical hazards. You will need to look carefully though, for they are often hidden from view within the tree canopy.

People & Pets

Tree work is noisy and exciting work that often attracts spectators, along with their pets, to the job site. Though it may be fun for them to watch you work, they should be considered a significant hazard to work operations. The behavior of people and their pets is unpredictable, such as a dog running into the work area followed by its owner. What's more, neither are protected from work hazards. Before you begin any tree work, scope out the work area to make sure no people are in the vicinity and warn and advise the ones who are (see page 83 for more people precautions).

Weather

Wind is one of the most potentially hazardous weather factors that affects tree work. During tree felling operations, even a light wind blowing from an unfavorable direction can cause the tree to fall off course or to "set back" on the chain saw bar trapping it in the back cut. Deciduous trees with leaves on (seasonal) and conifers (all year) are especially susceptible to this effect, because the canopy acts as a sail to catch the wind. Look for telltale signs of wind speed and direction by observing tree tops, flags, smoke from chimneys, or if near a lake, ripples and waves on the water. At the same time be watchful for signs of incoming bad weather. Especially in the west or northwest, listen for the sound of distant thunder and look for lightning or angry looking clouds developing. Have a plan.

PAY ATTENTION!—To Be Safe

TO BE SAFE in tree work requires careful and constant attention to one's surroundings. In a profession as dangerous as ours, attentiveness may just be *the* most important skill the tree worker can possess. This is true whether you're at the shop, on the road, or at the jobsite. The hazards you discover by being attentive, or miss from being inattentive, can mean life or death. Therefore, the importance of developing this skill cannot be overstated. Here are five suggestions that can help:

Be here now — An observant mind is an attentive mind—one focused on the present moment. Every time your thoughts wander from the task at hand—whether dwelling on a hurtful comment from your boss, planning the evening after work, or responding to the "ding" of a text message—you place extra demands on your finite attentional resources. This hinders your ability to notice the things which matter most in the here and now.

Use all your senses — Attentiveness involves using more than just your eyes. The ear may detect a hidden danger that could elude the eye (the buzz of hornets); the nose may discover a potential hazard that the eyes and ears could never detect (the stink of a leaky propane tank). To be fully attentive to job site hazards, it only makes sense to use all your senses.

Check your mood — Research has shown that when we are in a foul mood, we see less than when we are happy—our brain actually takes in less information from its surroundings. To cultivate a good mood, start with the basics: get plenty of sleep, eat healthy, drink plenty of water, hang out with positive people, and limit your time on social media.

Ask questions — Questions demand answers, even when you ask them to yourself. As you hunt for hazards at the job site, or while driving to and from job site, ask these three questions: "What do I see?," "What does it mean?," and "What should I do about it?" Questions like these help stimulate your search for answers and your mind for action.

Use checklists — It may seem obvious, but when you know what you're looking for, you're more likely to find it. That's where checklists come in. Checklists, like those found throughout this book, help you to be selective with your attention. They also help you remember what it is you are supposed to be doing and looking for, namely, hazards and obstacles to avoid.

**Turn to page 100 for Part 2 of "PAY ATTENTION!"*

Obstacles

As stated previously, an obstacle is any object you wish to avoid damaging (a target you want to miss hitting) or is a hindrance to safe work operations. Some obstacles can simply be moved to avoid being hit (page 84). Others may be successfully protected by covering them (page 85). Yet others may have to be avoided completely and subsequent action taken to do so (page 86). Be on the lookout for any of the following potential obstacles:

- ☐ ***Moveable obstacles***—cars, boats, trailers, lawn furniture, picnic tables, lawn ornaments, bird feeders, thermometers, grills, signs, bikes, etc.
- ☐ ***Immovable obstacles***—houses, sheds, decks, patios, lawns, gardens, fences, dog kennels, play areas, lighting, underground sprinklers, propane tanks, septic systems, overhead lines, sidewalks, and driveways.

Also, be alert to any objects which may already be broken or damaged, lest the homeowner hold you and your crew responsible. If you do happen to notice any preexisting property damage, be sure to bring it to the attention of your crew leader or the customer. If the customer isn't available, it's a good idea to take pictures to verify the damage.

People and pets, electrical hazards, ground hazards, wind, and obstacles are all present at this job site. What precautions would you take to make this site safe?

The Hazardous Nature of Tree Work

THERE ARE MANY WAYS the tree worker can get injured or killed at work—we've already looked at a few. Being aware of the causes and learning how to avoid them is an important step of job preparation and accident prevention. What follows are descriptions of the most common tree work related hazards and the injuries and fatalities they can cause.

Chain saws — The range of injuries resulting from the improper use of chain saws is extensive, including lacerations, avulsions, burns, vibration injury, and amputations. Survivors often receive deep, jagged cuts contaminated with bar oil and wood chips that require hundreds of stitches. Chain saw kickback and coasting saw chains are only a couple of the causes. As a ground worker, it is imperative you learn how to properly operate and maintain this potentially dangerous tool!

"Struck-bys" — This is a trade term that describes an injury resulting from a forceful impact by an object. For the ground worker, these objects typically include tree limbs and debris, trunk wood (sometimes the tree itself), and possibly even objects dropped by the climber. Dead and lodged limbs located in the canopy (known as hangers or widow makers) commonly become dislodged during felling operations and fall on or back toward the worker. Hung trees, falling trees, butt rebounds, and spring poles (see page 140) also account for many struck-by injuries and deaths.

The deadly spring pole

Struck-bys are consistently one of the Big Three causes of tree worker injury and fatality. (Falls from trees, buckets, and ladders, and exposure to electrical hazards are the other two.) Fortunately, they are also preventable by: (1) wearing proper personal protective equipment, (2) maintaining good communication between coworkers, (3) properly safe-guarding the job site using cones, flags, signs etc., (4) staying out of the drop zone of the tree until you've been told "all clear!"

Electricity — Exposure to electrical hazards is another of the Big Three causes of tree worker injury and fatality. All arborists, including ground workers, must be trained to recognize and avoid these hazards.

***Slips, trips, and falls* —** Accidents and injuries often occur when you lose your footing on uneven, steep, or slippery ground and steps. Ground surfaces littered with tree debris or covered with wet leaves, ice, snow, and mud all contribute to slips, trips, and falls. The best lines of defense from these injuries are to wear sturdy work boots that provide good traction in a variety of weather conditions, and to be attentive to where the hazards are, moving them out of the way whenever possible.

***Lifting injuries* —** One of the most common injuries for tree workers is back injury caused by lifting or moving heavy logs without using proper lifting technique (see page 148) or equipment. Make a habit of using the labor-saving tools described on pages 153-155, such as cant hooks and log dollies, instead of relying on brute strength to lift or move heavy objects. Other sensible options include cutting the wood smaller and asking a coworker for help.

Teamwork and labor-saving tools make moving a heavy log easier and safer along this common slip, trip, and fall hazard.

***Fatigue* —** Tree work is a physically and mentally demanding activity that can quickly lead to fatigue and exhaustion and then injury. Fatigue can cause you to react more slowly, ignore safety precautions, and become careless and reckless. To help prevent or offset the effects of fatigue: pace yourself, stay hydrated, and rest frequently—when it is appropriate to do so.

***Cuts, abrasions, and splinters* —** Wrestling limbs and logs is an inevitable part of tree work that leads to cuts, scratches, pokes, and punctures. Unprotected flesh is especially vulnerable to repeated contact with tree limbs, rough bark, and sharp wood surfaces. Wearing sturdy work gloves and long-sleeved shirts offers the best protection.

***Eye injuries* —** Eye injuries range from severe eye irritation and scratched eye tissue to blindness or eye loss due to airborne sawdust particles, wood chips, and tree branches. Prevention simply involves wearing protective safety glasses and safety helmets equipped with face screens.

Poisonous plants — Poison ivy, poison sumac, and poison oak are the three poisonous plants with which you may come into contact. All three contain urushiol oil, which can cause a skin rash or an allergic reaction. This condition affects about one-half the U.S. population. Symptoms range from mild and irritating to severe and life threatening. Be able to identify the plants that are located in your area and avoid touching them. Wearing work gloves will help minimize skin contact. If contact is made, wash your clothes in hot water and your skin thoroughly in as warm a water you can tolerate.

Insects, spiders, and ticks — Expect stings and bites from a variety of species of hornets, wasps, bees, spiders, wood ticks, and other similar pests, anytime you are working around grass, brush, and trees. Some of these pests have the potential to cause debilitating and potentially life-threatening conditions, such as Lyme disease (from deer tick bites), or anaphylactic shock (from hornet and bee stings).

Look and listen for hornets and have a plan to deal with them.

Refer to a good first aid book to learn how to treat injuries and complications that can result from insect and spider bites. The preventative measures common to them all are, 1) learn which of these pests inhabit the area you'll be working in, 2) look for their presence and avoid them, 3) wear appropriate clothing to discourage them, 4) apply bug-repellent to deter them, 5) use sprays to kill them, and 6) use first aid medicine when you are stung or bitten by them.

Weather — The weather conditions that are the most hazardous to tree workers are extreme air temperatures of heat or cold and snowy or rainy weather. The former can cause adverse health problems such as heat exhaustion, heat cramps, heat stroke, and frostbite; the latter contributes to dangerous slips and falls. Hypothermia, which occurs when body temperature falls below 95 degrees F, is another weather-related threat in cold or even mild temperatures (60 degrees F). Once again, refer to a good first aid book or take first aid training to learn how to recognize, prevent, and treat the symptoms of these weather-related health hazards.

Preparing for Emergencies

EVEN WHEN YOU HAVE TAKEN extreme safety precautions, accidents still happen. Ultimately they are out of our control. But one thing you can do is prepare for them *before* they happen. You can begin by addressing the following questions:

- Is there a fully stocked and labeled first aid kit available on the work truck? Where is it located?
- How close is the nearest person to offer help? Are they at home?
- Where is the nearest phone? If using a cell phone, is there adequate reception at your work location? Do the others know how to operate it?
- Who will you call for help in an emergency? Are emergency phone numbers posted in the first aid kit or work vehicle?
- Can you give the address and clear directions to your work location when calling for help? Consult the work order if you can't.
- Are you capable of providing basic first aid and CPR to a coworker? Seek the appropriate training if you can't.

Responding to Emergencies

TIME IS OF THE ESSENCE when an emergency occurs. How well you respond will depend in part on how much training you have had. But even those with little or no first aid training at all can still help out in any of the following ways:

- Make the area safe if a hazard exists.
- Call 9-1-1, or enlist a coworker to make the call, providing key information such as job location, number of injuries, nature of injuries, presence of hazards. Stay on the phone to provide more information or receive instructions if necessary.
- Help wave down and direct the emergency vehicle(s) to the job site and the injured party.
- Help with providing first aid by following the instructions of those who do have training.

CREATING A WORK PLAN

A tendency of many work crews is to jump into action too soon. While they may have read over the work order and inspected the job site for hazards and obstacles, they often begin work before organizing that information into a work plan. And without a work plan, work is more likely to be performed in a confused, haphazard, and unorganized fashion—a perfect recipe for making mistakes and causing accidents.

As eager as you and your crew may be to begin working, it is absolutely essential that a work plan be created first. While it is the crew leader who typically oversees the process, everyone on the work team potentially has something helpful to contribute—insights into work efficiency that other workers might not have, for example. A good work plan will address the following issues:

1. ***What work is to be done***—The work order is the starting point for creating a work plan. Make sure that you read through the work order and understand what work is to be done, especially on the trees you'll be working on. Any information from the work order that could affect work operations—special tools needed, where to park vehicles and equipment, etc.—should be included in the work plan as well. If you don't understand something about the work, ask your supervisor for clarity. Or, if you notice something unexpected or unmentioned on the work order, bring that also to his or her attention.

2. ***What site hazards exist and how they will be avoided***—Everyone on the work crew needs to be aware of the hazards and obstacles found during the job site inspection. The work plan should address what they are, where they are located, and what measures will be taken to deal with them. In particular, protecting people and property should always be your highest priority. The presence of hazards and obstacles will also affect your choice of work methods and equipment.

3. ***What work methods are to be used***—There are many ways to get the job done successfully. However, one of the goals of creating a good work plan is deciding which methods are *most* effective. That is, the methods which are most suitable and most efficient for the job at hand. And the more tricks of the trade your crew knows, the bigger the play book.

4. ***What equipment is needed***—The work to be done (#1), the presence of hazards (#2), and the methods that will be used (#3), are all factors in determining which tools will be needed. If, for example, the plan is to rope down limbs to avoid an obstruction, then it's likely you'll need a rigging line and lowering device. If time at the shop was spent wisely, these tools will already be on the work truck and in good repair.
5. ***What the job assignments are***—The work plan should be crystal clear on job assignments: who will do what task, at what time, at what location, and in what manner. Furthermore, everyone should understand what the other workers' responsibilities are, as well as where each person will be located at the job site while performing them.

Explaining the Work Plan (The Job Briefing)

Simply put, the job briefing is a time when the work plan is explained to the work team by the arborist in charge. But explanation is not enough if the details of the plan fall on confused minds or deaf ears. In other words, the work plan needs to be communicated in such a way that it is thoroughly understood by everyone and not misunderstood by anyone. As a ground worker, you can't control how the plan is communicated, but you can take responsibility to ask clarifying questions so that you understand what you need to do. You might even ask a clarifying question for another worker who may be too shy to speak up for him or herself.

Changing the Work Plan

> "Everyone has a plan until they get punched in the mouth."
>
> —Mike Tyson (boxer)

To be sure: create a plan and explain the plan, just don't fall in love with it. Once work begins unexpected circumstances often develop, which may require making changes to the original work plan. This can happen for a number of reasons: a hazard or obstacle is discovered that went undetected during the assessment process, the customer requests a change to the work order, additional employees show up on the job, or there is a change in weather conditions. Sometimes plans change simply because the original plan wasn't working well. Whatever the cause, it's important that all team members are aware of the revised work plan before proceeding.

A Job Briefing From 1812

THERE ARE A NUMBER of similarities between the battlefield of a soldier and the job site of the tree worker. The most notable similarity is that both are places of ever-present danger where good communication among the work crew is essential to the success of the operation. And good communication begins with briefing the crew about the war plan, or in our case, the work plan. Sometimes hearing an account of an activity similar to our own, but in a different context, brings fresh insights and appreciation to our own application. Therefore, consider the following "job briefing" given by Captain Jack Aubrey to his crew aboard the war ship *Surprise*. It includes all the elements of a good work plan as described on the previous two pages.

> "Right lads, now, I know there's not a faint heart among you, and I know you're as anxious as I am to get into close action. But we must bring them right up beside us before we spring this trap. That will test our nerve, and discipline will count just as much as courage. The *Acheron* is a tough nut to crack—more than twice our guns, more than twice our numbers, and they will sell their lives dearly.
>
> Topmen, your handling of the sheets to be lubberly and un-navy like. Until the signal calls, you're to spill the wind from our sails, this will bring us almost to a complete stop. Gun crews, you must run out and tie down in double quick time. With the rear wheels removed, you've gained elevation, but without recoil, there'll be no chance for re-load, so gun captains, that gives you one shot from the larboard battery—one shot only. You'll fire for her mainmast. Much will depend on your accuracy, however, even crippled, she will still be dangerous, like a wounded beast.
>
> Captain Howard and the marines will sweep their weather deck with swivel gun and musket fire from the tops. They'll try and even the odds for us before we board. They mean to take us as a prize. And we are worth more to them undamaged. Their greed will be their downfall. England is under threat of invasion, and though we be on the far side of the world, this ship is our home. This ship *is* England. So it's every hand to his rope or gun, quick's the word and sharp's the action. After all, *Surprise* is on our side."
>
> —from the movie, *Master and Commander*[12]

PROTECTING PEOPLE

After the job briefing your first priority is to prepare the job site by making it safe for people—yourself, workers, bystanders, pedestrians, and vehicular traffic. You already began doing so when you first arrived at the job site and took measures to control traffic and connect with the customer. The focus now is to keep people from walking into the *work zone.* The work zone (or work area) is a defined area at the job site where tree work is being performed and where potential hazards exist. One effective way of doing this is to set up a barrier around the perimeter of the work area using bright flagging, safety cones, or signs. In addition to these safety precautions, it may be necessary to post one or more tree workers in strategic locations once the work begins. This will help ensure no one accidentally enters the area at a crucial moment.

A location that is safe for people to stand will depend on the type of work being performed—pruning, technical rigging for removal, or felling. But keeping bystanders at least two tree lengths away from the tree is generally a good rule of thumb for their safety. It would be wise to increase this distance when felling large dead trees to avoid injury from the "shrapnel effect" or flying debris that occurs when the tree crashes to the ground.

The measures taken to protect the tree workers themselves will also depend on the type of work being performed. This is discussed in more detail on pages 115, 117, 131, 134.

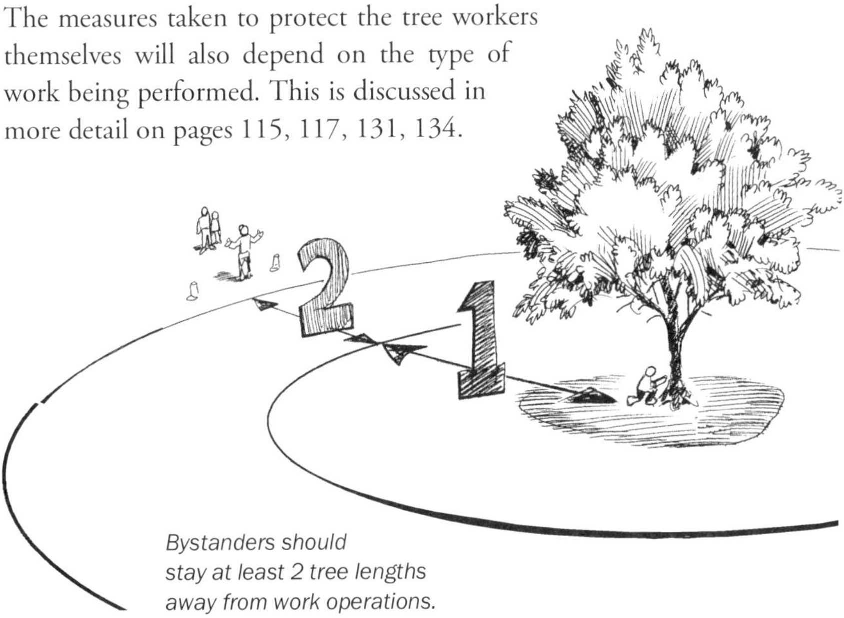

Bystanders should stay at least 2 tree lengths away from work operations.

PROTECTING PROPERTY

The final step of preparation before beginning work is to protect property from being damaged—the customer's, as well as affected neighboring properties. The term "property" in this discussion refers to both objects and structures of value as well as the land or grounds. The property you will be most concerned about protecting are the objects/obstacles you noted during the job site inspection.

Important as this aspect of work preparation is, protecting property is frequently either poorly executed or neglected entirely. Granted, it takes time (and sometimes ingenuity) to adequately protect property. But when you consider the cost of time and money to repair or replace damaged property, it is time well spent. What follows are three basic ways to protect property from damage: move it, cover it, or avoid it.

1. Move It

The easiest way to protect something from getting damaged is to move it out of the way. Be realistic; things don't always go as planned, so move even those objects you do not think will be in the way. Refer to the list on page 75 of potential obstacles that could possibly be moved.

Move it or lose it! And then afterward, put it back where you found it.

Some obstacles, such as fences, may be dismantled and reconstructed after felling is finished. You may want to cover the holes left from the posts so they can be reinstalled in the same location afterwards. This is a good trick for any object you remove from the ground: flag poles, bird feeders, or signs. Safety cones work great for this purpose.

Obstacles such as overhead power and communication lines can be moved or dropped by contacting the appropriate utility company. Finally, do not overlook moving your own gear. Place it in a safe and visible location.

2. Cover It

If you can't move something to avoid damage consider how you might protect it by covering it instead. Four excellent ways of doing this are by using plywood, tarps, cones, and tires. Plywood is great for covering walkways, lawns, decks, windows and roofs as it offers protection from limbs and tree debris during felling and rigging operations. Hinged plywood propped up over plants and garden areas is another practical way of protecting property.

Large tarps can also be used to cover the same items along with a variety of other objects, though the protection it offers is considerably less. Tarps are also useful during cleanup when used to drag out brush and debris and when used to protect lawns from sawdust accumulation while bucking wood.

Bright orange safety cones are very useful for covering or identifying smaller objects you wish to protect, as well as their intended purpose of warning pedestrians and drivers of vehicles that work is being performed in the immediate area. Objects that are most commonly protected with cones are small plants, young trees, sprinkler heads, and landscape lighting.

Finally, stacked car tires (without the rims) make a great cushion for tree limbs and trunk sections to land on in order to protect sidewalks, lawns, and other landscape features. You may also use cut log sections from a previously felled tree for the same purpose. When using log sections to protect sidewalks for instance, place one on each side of it and not directly on it.

3. Avoid It

If you can't move an obstacle or adequately protect it by covering it, which is almost always the case for such obstacles as driveways, homes, sheds, and septic systems, then you are left with no alternative but to avoid it completely. During tree removal and pruning operations, limbs and tree sections will most likely need to be lowered with a rigging line to avoid any property damage. Whatever the scenario is, coming up with creative solutions to avoid obstacles is part of the challenge and fun of tree work.

Estimating Tree Height to Avoid Hitting Obstacles

It's usually quite obvious which objects need to be moved, protected, or avoided—they're the ones too close to the action. But it is less obvious when the obstacle is some distance from the tree, as is the case during tree felling operations. The only way to know for sure if the tree will reach the obstacle is to estimate the tree's height.

All you need to accurately do this is your arm, your eyes, and a straight stick. This method is referred to by many as the stick method or "stick trick." Lumberjacks have been successfully using this method for decades; yet, many tree workers are unfamiliar with this procedure. Most customers who observe a worker performing this trick usually react with amazement when they see the accuracy of the results.

Using the Stick Trick

1. Begin by making a gauge stick approximately 32 inches long or a little longer than your outstretched arm. This may be a straight limb found at the felling site. Stand in a position, in line with where you anticipate the tree to land. Grip the stick so the distance from your eye to your outstretched hand is the same as from your hand to the top of the stick.
2. Hold the stick vertically so that it forms a 90-degree angle to your arm, which should be extended horizontally (level). It is important that the stick be held perfectly vertical with your arm level and fully extended.
3. Sight the tree by aligning the bottom of the gauge stick (or the top of your hand) with a spot on the trunk the same distance off the ground as is your hand. While keeping your head stationary, move forward or backward until the tree appears to be the same height as the stick. After adjusting for the difference between your arm height and the

height of the felling cut (4 feet in the illustration below) you are standing at a distance equal to the tree's height and at the spot where the top will land when felled.

It's not a bad idea to get a second opinion from a coworker. Any woodworker knows that it's a wise practice to "measure twice, cut once." This is just as relevant for someone felling trees. Ideally, you should always allow extra space beyond where the top is expected to hit to allow for error in height estimation and for broken tree limbs being tossed in the fall.

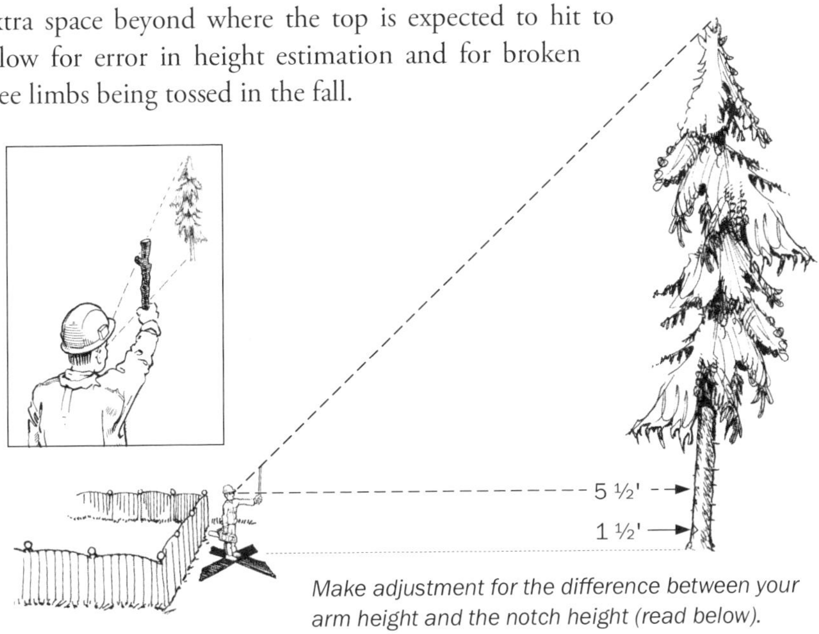

Make adjustment for the difference between your arm height and the notch height (read below).

The accuracy of this method is maximized when measuring trees that are vertical and where the ground is level. Make adjustments for the difference of your height (at eye level) and the height of the felling cut (as shown above). For example, if the height to your eye level is 5½ feet from the ground and the height to the notch is 1½ feet (a 4-foot difference), move backward 4 feet (after measuring) to compensate for the height difference.

You must also compensate for trees with a back or forward lean, as they will appear shorter than they really are. For every foot of back or forward lean the tree has, add at least an extra foot to your height estimation. When estimating tree height on sloping ground making grade adjustments gets a bit more tricky. Know this at least: if you are standing above the tree base you will need to add distance; below the tree base subtract distance.

PART 2: JOB EXECUTION

Up to this point in the chapter, you have inspected the job site, taken a number of measures to protect people and property, and had the work plan explained during the job briefing. Now it's time to begin performing the work you were hired to do. It's time to work the plan. And to do that you'll need to learn the practical skills and knowledge of production tree work—the focus of the remainder of this chapter.

There are no shortcuts to your education. It will take lots of practice and persistence to become competent. Quite likely, it will also cost you some blood, most certainly much sweat, and maybe even some tears; you might even have thoughts of quitting. However, if you hang in there long enough, it will all be worth it as you experience the pleasure of acquiring the skills of the trade and putting them to use to perform a job well done.

SELECTING & PREPARING EQUIPMENT

From the job briefing you will have learned which tools and equipment are needed for the job. Assuming that the equipment is ready to go (after being serviced at the shop), it's just a matter of getting the gear from the work truck to the location where it will be needed at the job site.

In addition to this equipment and gear, a rescue kit should also be brought from the vehicle and made readily available near where the work is being performed. This kit includes all the necessary climbing gear and emergency equipment needed to conduct an aerial rescue and apply first aid. Many climbers often install a second climbing line in the tree as well. Commonly called an access line, this is helpful both for work efficiency and to reach a climber should an aerial rescue need to be performed. As a ground worker you may be asked to install such a line (see pages 101-107) and possibly help with a rescue from the ground.

Place selected equipment close to where the work is being performed, but not so close that it could get damaged from falling limbs, etc. A common practice with many work crews is to set up a "base camp" for the gear. This is simply a centrally located area at the job site where most of the work equipment is stored so it can be easily seen and accessed. Some of the gear can be propped up against a tree or building. Other equipment, such as ropes, pulleys, slings, or other loose items, might be placed on a tarp laid out on the ground to provide protection from dirt, water, or snow. Knowing where they are also helps prevent leaving any tools behind when packing up at the end of the work day. Of course, any equipment that can be easily seen by the work team can also be easily seen by others. If you are working in an area where gear can be stolen, you will need to keep a careful lookout and take precautions such as locking vehicle side bins.

A separate area should be set up for chain saw refueling and servicing that is at least ten feet away from where they will be used. Avoid refueling saws near work tools and equipment and on property features such as grass or driveways. Instead, refuel saws on a tarp placed on the ground or on the tailgate of a work truck, either of which will protect equipment and property against gas and oil spills.

When certain tools and equipment are no longer needed on the job site, bring them back to the work truck and stow them away in their proper place. There will be that much less gear to locate and put back later. In your eagerness to be efficient, just make sure you don't put away gear that might yet be needed.

Having a "base camp" for tools makes them easier to locate.

Never place equipment on top, under, behind, or in front of the work truck or chipper, no matter how visible it might appear to you. You never know when another worker, unaware of the gear near it, might jump in the truck to move it. For the same reason, always shut the doors on the work truck's side/tool bins after removing or stowing gear. If any gear does get damaged or malfunctions, report it to your supervisor as soon as possible. At the very least, make a note about it on the work order to remind yourself to mention it at a more suitable time.

HELPING THE CLIMBER

One of the ground worker's most important responsibilities at the job site is to help the climber. Your assistance will range from helping set up and steady ladders to "running the ropes" and being the climber's extra pair of eyes. Whatever the task, your help is vital to the success of the job and valued by the climber aloft.

Setting Up & Supporting a Ladder

Ladders are most commonly used in two ways at the job site. First, as a means for a climber to get into a tree; second, for a worker to get onto the roof of a house or other structure. As a ground worker, you will be expected to know how to safely set up a ladder for either application, as well as steady the ladder when one is being climbed by another worker.

Proper Ladder Setup

The ideal ladder angle is about 75 degrees. This can be attained by applying the 4:1 (height-to-lean) rule. For example, if the ladder is 24 feet high at the support point (tree limb, trunk, roof edge), the base of the ladder should be 6 feet off plumb (one-fourth the height). A simple way to attain the same results is by using the "outreach method" as described below:

1. Set the ladder into position against a tree limb or eaves of a structure. Secure the locks on extension ladders.
2. While standing straight, place your toes against the side rails at the bottom of the ladder.
3. With arms extended straight out and parallel to the ground, the hands should be able to touch the side rails of the ladder.

Proper ladder angle

Ladder Safety (In trees)

1. A ground worker should support the ladder from below while the climber ascends. Remove the ladder when the climber is safely secured in the tree to avoid damage from dropped limbs. When convenient, the ladder can be stowed on the truck if it is no longer being used.
2. Sometimes a ladder will be more secure when positioned against a side branch instead of the main trunk of the tree, where it can tip from side to side against the top rung.
3. Since climbers should not be carrying tools or material in their hands while climbing the ladder, a ground worker must be available to send them up on request.
4. Ladders made of metal or other conductive material must never be used near an existing electrical hazard.

Belaying the Climber

The ground worker may also be asked to belay the climber as he or she body thrusts or free climbs into the tree. There are two ways this can be done. First, after the climbing line has been installed and the climber has been secured to the line, the ground worker can provide a belay using either a figure-eight descender or Münter hitch (both shown below) attached to an anchor at the base of the tree. The ground worker takes in the slack as the climber ascends.

A second, less advanced and simpler method of belay is for the climber to tie in to his or her climbing system using an approved friction hitch/climbing knot with a micropulley slack tender before leaving the ground. With this method, all the ground worker needs to do is pull down on the climbing line. This will cause the micro pulley to advance the friction hitch. If desired, the climber may perform a self belay on the line at any point during the ascent.

Know Your Knots: Part 1

FOR MOST NEW TREE WORKERS, tying knots is one of the most difficult and maddening skills to learn. But learn them they must, for knots are one of the most frequently used tools in tree work. Fortunately, you don't need to know lots of them to be a good ground worker (not initially anyway). But you will need to know about ten knots very well (the ones listed below). This really isn't many considering there are over a thousand different knots. To *know* your knots means selecting the one most appropriate to the situation and tying it *exactly right*—every time. As knot expert Clifford Ashley famously stated, "a knot is never nearly right; it is either exactly right or it is hopelessly wrong."

How to tie, dress, & set a knot — Tying a knot exactly right means more than just following "the procedure," like making the "rabbit come out of the hole, go around the tree, and back in the hole" as with the bowline knot (as important as this step is). But tying it right also means the knot must be properly finished by ***dressing*** and ***setting*** it.

Dressing a knot means to properly align or arrange all the parts of the knot so that it matches the descriptions and illustrations in the book. Improperly dressed knots can result in significant strength loss.

Setting a knot involves tightening all its parts so they properly touch, grab, and press against each other. This creates friction on the rope—the reason a knot works. An improperly set knot can "capsize" (distort) into something completely different, or nothing at all; it can even come untied completely after a load is applied, and can also become impossible to untie afterward, often referred to as a knife knot.

Must know knots — These are the knots every ground worker should learn first, for they are the most frequently used, versatile in application (except for the quick hitch), and foundational for learning other knots later on.

1. Quick hitch: page 94
2. Midline clove hitch: page 95
3. Slip knot: page 96
4. Figure-8 knot: page 101
5. Pile hitch: page 103
6. Bowline: page 107
7. Running bowline: page 107
8. Timber hitch: page 112
9. Butterfly knot: page 135
10. Prusik hitch: page 136

**After you've mastered these, turn to page 172 to learn your next two knots.*

Know Your Knots: Part 2

TO BECOME A TRUE "KNOT HEAD" you'll need to learn the language. First, to understand what other knot tyers are talking about and secondly, to understand the tying instructions in this book.

Knot Categories

Knot — A general term referring to all knots, hitches, and bends.

Bend — A knot that joins two rope, cord, or webbing ends together.

Hitch — A knot that secures rope or webbing to an object or the ropes own standing part.

Loop knot — A knot with a fixed loop used for attaching to an object.

Stopper knot — A knot tied in a rope to "stop" the rope from pulling through another object such as a climbing hitch or pulley.

Rope Parts

Bight — Any well-defined, doubled section of rope that doesn't cross itself.

Loop — A turn or bight of rope that crosses itself.

Running (or bitter) end — The free end of the rope or the end not being used to tie or rig with.

Standing part — All the inactive parts of the rope uninvolved with rigging or knots.

Round turn — Two loops of rope around an object. A ***crossed round turn*** is made by passing the round over the second (used when tying a clove hitch).

Turn — One loop of rope passing around an object.

Working end — The active end of the rope used to rig or tie off to something.

Turn to page 137 for Part 3.

Rope Anatomy

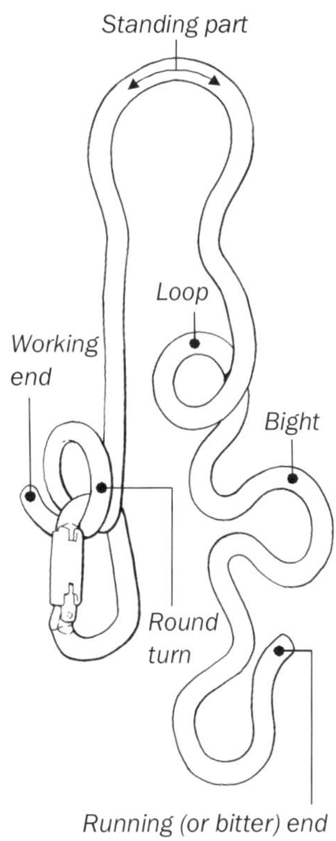

Sending Gear up to a Climber

One of the marks of a capable ground worker is demonstrated by how efficiently he or she secures a requested piece of equipment to the climber's line. Not in some haphazard sort of way, but in a manner which makes it easy to remove from the line. What follows are things climbers most commonly request to have sent aloft (and send down to a ground person) and the best knots for securing them to their line.

Another rope—This is probably the most commonly requested piece of equipment by a climber aloft. As easy as it sounds, securing another rope to the climbing line can be tricky, which is made evident by the attempts of most rookie tree workers. While there are several good knots for this application, such as the sheet bend and attachment bowline (see figure 1 on the bottom of page 107), by far the best one is the ***quick hitch***. While this may be the only use for this knot, it's one worth knowing as it lives up to its name as a knot quick to tie and untie. A quick yank on the tail and the two ropes instantly separate. This knot is sometimes challenging for beginners to learn, partly because they are learning so many other knots at the same time. The climber, however, will be especially grateful that you do. In fact, this should be one of the first knots for you to master.

MUST KNOW KNOT: Quick Hitch

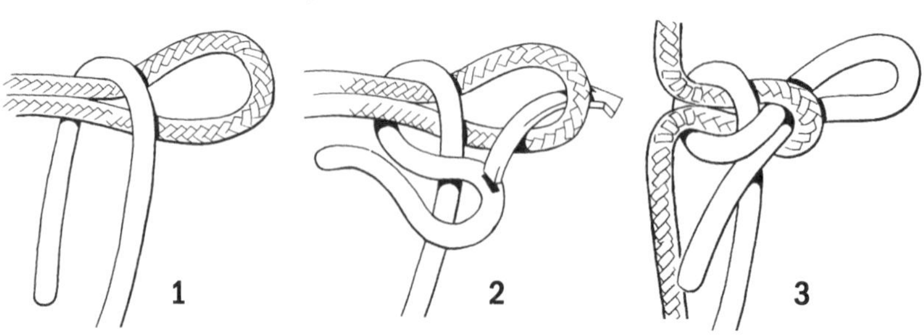

How to tie: *(1) Pass the end of an attachment rope (clear rope) around a* **bight** *of the climber's line. (2) Pass a bight of the attachment rope in front of the standing part and through the bight of the climber's line. Leave a suitable length of tail. (3) Draw the hitch up tightly by pulling on the bight of the attached rope and the climber's line. To release the hitch and separate the two lines, simply pull on the end of the attached rope.*

Note: *The quick hitch can also be used for attaching a rigging sling to the climber's line.*

Hand tools—The most frequently requested hand tools to be sent up to a climber are a pole saw, pole pruner, hand saw, cabling tool (e.g., bolt cutter, come-along), and in some instances, even climbing spurs. The ***clove hitch*** (sometimes backed up with a half hitch) is the most commonly used knot for securing these tools to the line. When sending up a pole saw, pole pruner, or hand saw (the latter of which is usually dropped by the climber), make sure the cutting edge is oriented away from the climbing line. It may be necessary for a ground worker to guide the tool away from interfering limbs as it is raised in the tree by the climber.

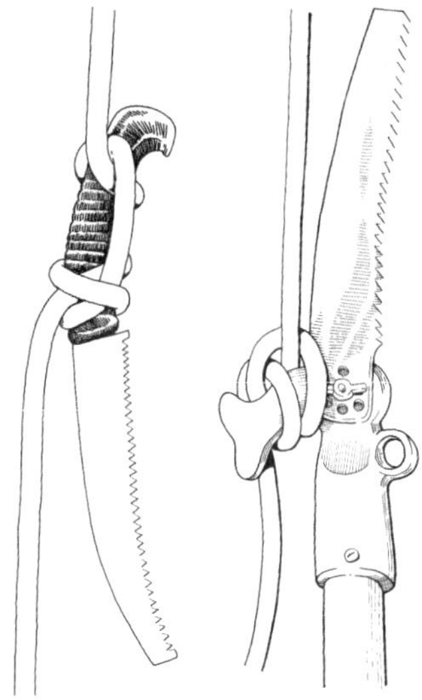

Clove hitch (both), with half hitch (hand saw only).

Water bottle or thermos—A good ground worker will not only remind the climber to drink water to stay hydrated, but will also know how to send it up when it's needed. The most commonly used knot for securing a water bottle to a line is the midline clove hitch. Adding an optional half hitch above the knot (as with the hand saw shown above) will help keep the bottle upright, making it less likely to catch on limbs during retrieval.

MUST KNOW KNOT: **Midline Clove Hitch**

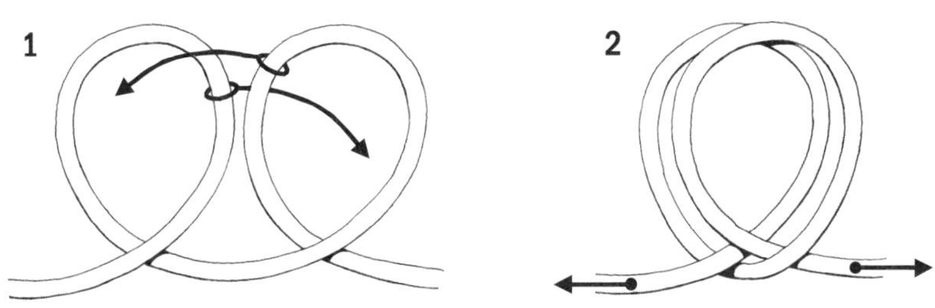

How to tie: *(1) Form two loops in the rope with opposite twists. Pass the right loop in front of the left. (2) Place the two loops around the object and draw up tight by pulling on the two standing parts of the line.*

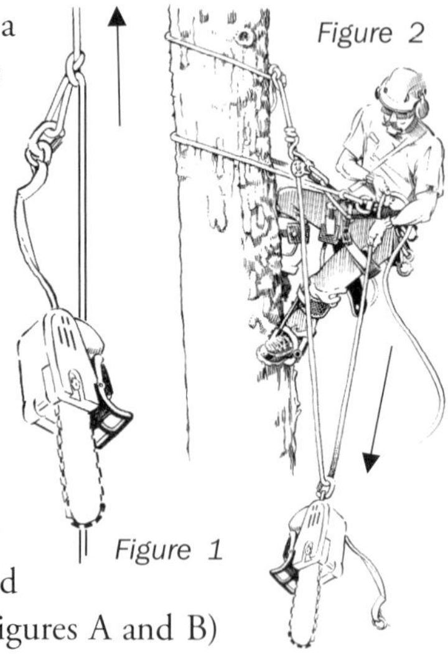

Chain Saw—The easiest way to send a chain saw aloft (and remove it from the line afterward) is by clipping a biner to the chain saw lanyard (or just the saw itself), and then to a ***slip knot*** tied in the climber's line (figure 1 right). (Better yet, attach the saw to the end of an installed rigging line and pull it up for the climber.) Use the tying method shown below in figures 1 and 2 for lighter weight saws. But if the saw is particularly heavy and is being tied near the end of the rope, you should consider tying the slip knot in reverse (figures A and B) to prevent the end of the rope from pulling through the knot.

Always apply the chain brake before sending the saw back to the climber. The climber may also ask that the saw be warmed up as well, but shut off afterward. It may be necessary to guide the saw away from interfering limbs as the climber hoists it aloft. Also, avoid letting the tip of the saw bar make contact with the ground to prevent dulling the chain.

The easiest way for a climber to *lower* a saw is to clip the saw to the climbing line in the bight of rope that is formed as the climber lowers the running end of the line, hand-over-hand (figure 2 above).

MUST KNOW KNOT: **Slip Knot**

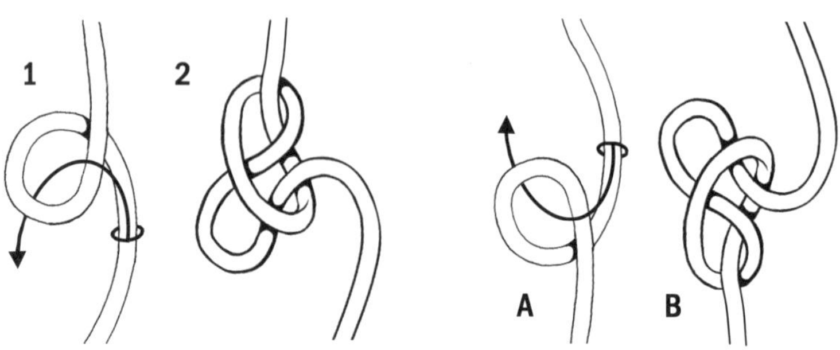

Note: *A slip knot is simply a slipped version of an overhand knot. It is intended to be loaded from only one direction, depending on the application.*

Maintaining Good Communication With the Climber

Poor communication between workers, especially between climbers and the ground crew, is the cause of many accidents and relational tensions that occur at the job site. To avoid either, it is imperative that you learn the following methods of good communication with the climber.

Use a voice command-and-response system—"Command and response" is the most widely accepted method of communication between the ground worker and climber. Simply put, a worker initiates a command but does not proceed until an audible response has been given and heard. For example, a ground worker who wants to retrieve limbs from under the tree alerts the climber with a verbal command of "Going under!" (or just "Under!"). But not until the climber acknowledges the command with a response of "OK!", or "All clear!" does the ground worker proceed and enter the drop zone. Even then, the ground worker should always take a final look at the climber before proceeding—just in case. Below are a few examples of how an audible command-and-response system can be used.

Command	***Response***	***Sample Application***
Stand clear! (climber)	All clear! (groundie)	A climber's warning before cutting a limb.
Headache! (climber)	Act fast! (groundie)	An urgent warning that something is already heading downward.
Slack! (climber)	Ok/Roger! (groundie)	A climber's request to release rope tension caused by ground debris.
Up rope! (groundie)	Thank you! (climber)	Alerting the climber when returning the rigging line.
Under! (groundie)	Ok/All clear! (climber)	A ground worker requesting to enter the drop zone.

Hand signals can also be an effective way of communicating, especially when things get noisy. A held up clenched fist typically means "Stop" or "Hold your position"; a thumbs up signals "All is well" or "I'm ready"; and a hand passing back and forth across your throat means "Shut off the ________" (chain saw, chipper, or other noisy equipment). It matters little what the signal is as long as everyone understands its meaning.

Use a whistle—Another important, but often overlooked method of communication is the use of a whistle. It is highly recommended that each worker have one. A whistle is a compact, inexpensive, and highly effective tool for interrupting work operations if necessary. It should be loud enough to carry above the noise of a running chain saw or chipper and accessible enough to use when you urgently need it.

Use your "outside voice!"—Whether you are giving a command or responding to one, it is important that your voice be loud enough for the climber to hear you. Don't expect a climber who is forty feet up a tree, with the wind blowing, the chain saw idling, and who may already be half deaf from years of tree work, to understand a weak half-hearted command or response from down under. Use your outside voice and be proud of it!

Always acknowledge—As a ground worker you'll be given all sorts of commands, instructions, and requests from the climber, many of which are non-critical in nature, but important nonetheless. But whatever is spoken from above always give some type of acknowledgement with words ("Gotcha!", "Roger that!", "Aye, aye, Captain!") or signals (thumbs up) that you heard and understood what the climber was saying. If no response is made on your part the climber will assume that the message wasn't heard and therefore, must repeat it and waste valuable time.

Use your words wisely and sparingly—Don't waste the climber's time with unnecessary comments or idle chatter irrelevant to the work at hand. Use your words wisely and sparingly. Should you alert the climber to unseen dangers? Always!; To a low flying eagle overhead? Maybe; To comments about birds in general? Save it for later. Words of encouragement ("you da man!—or woman!"), or questions of concern ("how are you doing?), when given at an appropriate time, are also appreciated by the climber. Just remember, the climber has enough to concentrate on than to consider whether what you're saying is important or not.

Ask permission—One of the most dangerous times in tree work is when the wood chipper is running while the climber is working in the tree. Verbal communication is impossible and a climber trying to get the groundies attention is difficult. That's why it is so important that the ground worker always ask permission before starting noisy equipment. Also, when you are running noisy equipment, take frequent visual checks on the climber.

Being the Climber's Extra Eyes

Another important way that ground workers help climbers is by being their extra pair of eyes. Your "boots on the ground" perspective enables you to see things the climber can't. Things such as dead limbs to be pruned, hung limbs or tools to be retrieved, and hazards and obstacles to be avoided (hornets nests, electrical hazards, windows), are just a few examples of what the climber needs to be reminded of and warned about. Never assume the climber is always aware of what's going on. There have been instances of climbers topping out a tree with their climbing system still connected (guilty!) and climbers descending on a short rope without a stopper knot tied in the end (see page 101).[13] If you're paying attention, you can warn them of such things. Their survival depends on it.

Why is this climber in danger? As his helper you need to know.

Anticipating the Climber's Needs

Anticipating the climber's needs by offering the help or tool before having to be asked for it is yet one more mark of a great groundie. Perhaps the climber will soon need a bigger chain saw to block down larger trunk wood. Well then, make sure the saw is fueled up and readily available *before* it's needed. Maybe a beefier block and rigging line will soon be in order. Again, be ahead of the game and have them ready (see *Don't Waste Your Trips* on page 152). Perhaps the climber's need is as simple as a drink of water. In which case the thoughtful question, "Do you want your water bottle sent up?" can be the much needed break and refreshment the climber needs but hadn't thought of until you mentioned it.

Whatever the need might be, however, it is important to maintain good communication and inform the climber of your intentions. And, if you must leave the immediate work area, ask if this is a good time to do so. Climbers should never have to wonder where you went off to. Nor should they have to wait an unnecessarily long time for you to get something. Put some spring in your step—run if necessary—to get what you need quickly.

PAY ATTENTION!—To Be Useful

THE NINETEENTH CENTURY preacher C.H. Spurgeon once said, "Let us go about the world with our ears and our eyes open, ready to avail ourselves of every occasion for doing good; let us not be content until we are useful, but make this the main design and ambition of our lives." This is a mouthful of good advice, especially in this age of mindless distraction and self-absorption. Simply put, Spurgeon means for us to be watchful for ways to be useful for the good of others; to "see a need and meet a need." Doing so makes work more productive and makes a positive difference in the life of others as well as our own. Here are some ideas to get you started:

To coworkers — We've already looked at a number of ways that ground workers help climbers, but what can you do for your coworker on the ground? Plenty. You could offer a helping hand on a pull line, the extra muscle needed to lift a heavy log, or even a refresher lesson with a difficult knot. To a weary coworker you might suggest a water break. In some cases, the most helpful thing you can do for a coworker is offer a word of praise—Mark Twain once said that he could live two months on a good compliment. It's true.

If you have difficulty finding ways to be useful, then simply ask, "What can I help you with?" Eventually, you'll get the hang of it. And it really doesn't matter how big or small the act of helpfulness is, for "the smallest good deed is always better than the grandest good intention."

To customers — And what about your customers? How might you be useful and helpful to them, especially in a way that is above and beyond what is already expected and required? You could offer to open a door for a customer struggling with a load of groceries, or maybe an explanation about a work procedure of which they seem interested in. Whatever the case, it is only when we pay attention that we can do something about it.

To the cause — Not all your efforts of usefulness need to be directed toward the good of others in order to be meaningful. Just being useful for the good of the job is also important. You may not get the pat on the back that sometimes comes from helping others, but doing the good you know you should do—without an audience—is its own reward. So, look around and ask yourself, "What can I do to help complete this job?" And then go and be useful and do it.

**Turn to page 118 for Part 3 of "PAY ATTENTION!"*

INSTALLING A ROPE IN THE TREE

Installing a rope in the tree with a throwline and throwbag is a skill all ground workers will eventually need to learn. It is an essential skill for setting rigging lines from the ground during natural-branch union (natural crotch) rigging and to set a pull line in the tree during felling operations. Those who become proficient with this skill are often asked by climbers to install their climbing line as well—a task they typically perform themselves. (One more item on your ever increasing "helping the climber" list.) Doing so frees them up to make preparations for climbing and other tasks.

The standard equipment for rope installation is a suitable length of throwline (120-150 feet minimum) and a shot filled throwing weight (8-16 ounces), also referred to as a shot pouch or a throwbag (my preference). If it's a climbing line you are installing, the climber will most likely point out the branch he or she would most prefer—likewise with a rigging line. Installing a pull line for felling purposes, however, is a different matter. Here, the biggest concern is the height at which the rope is installed in the tree. The higher in the tree it is secured, the more leverage you potentially gain to make it easier to pull the tree into the direction you desire. The ideal height for a pull line installation is about 2/3-3/4 the height of the tree. A rope installed lower than this will lessen the leveraging potential, but may still be satisfactory for small trees with only a slight back lean. To install a rope, follow the instructions on the following two pages.

MUST KNOW KNOT: **Figure-8 Knot**

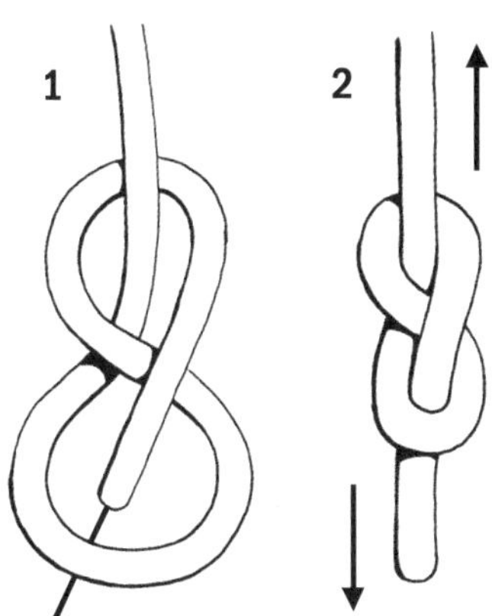

Knot uses:

- *The foundational knot for learning the slipped figure-8 knot used for attaching a throwline to a throwbag (page 102).*
- *As a stopper/safety knot tied in the end of a climbing line.*
- *Tied in the end of a rope to make it easier to find when stored in a rope bag or bucket.*

Basic Rope Installation Procedure

STEP 1

Tie the throwline to the throwbag with a slipped version of the ***figure-8 knot*** (see illustrations at right and page 101 for the basic figure-8 knot). Though there are several other knots that will also work for this job (see bowline—fig. 2 on page 107), this knot provides a secure attachment that unties quickly when you are ready to tie on the pull line after the throw (in step 3).

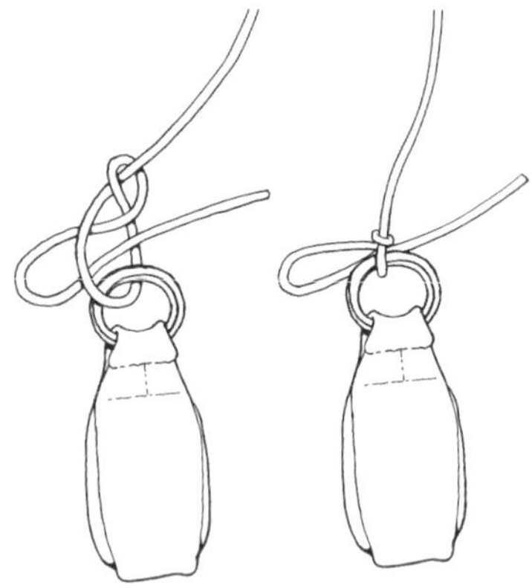

STEP 2

Aim for a suitable branch union or sturdy limb in the tree that can support the weight of the climber, rigging forces of lowering, or the forces required to pull a tree over during felling operations. Select wide, u-shaped branch unions whenever possible to allow the throwline (and attached rope, step 4) to pass through easily.

Throw the line and weight using the single-hand toss method (shown below) or another preferred method. Make sure the line is free of tangles, ground litter, and your own feet. Tying a slip knot 2-3 feet up on the line helps you grip and release the line better. The knot can be easily pulled out afterward. Alert others nearby that you are about to throw by shouting, "Stand clear!"

Throwing tip: *Your throwing accuracy can be greatly improved simply by focusing on the target branch union and visualizing the throwbag and line passing over it.*

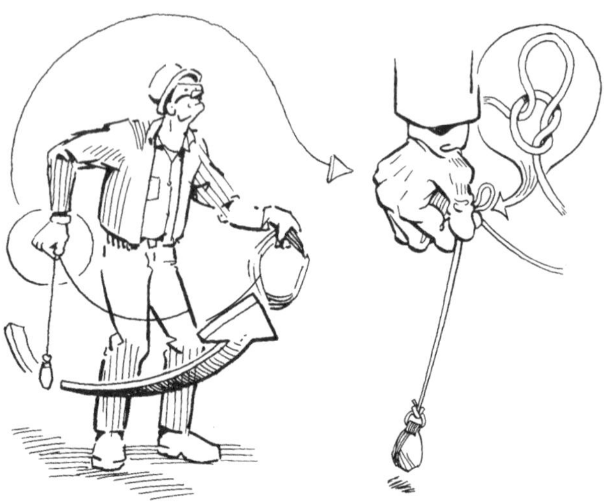

STEP 3

Allow the bag to fall back to the ground after a successful shot has been made. It may be necessary to first manipulate the bag over interfering limbs so the throw line runs nearly vertically along the back of the trunk.

Remove the throwbag from the line and attach the climbing, rigging, or pull line using a ***pile hitch*** (see below). For a more streamlined effect to get through narrow branch unions add a series of half hitches finished with a girth or clove hitch (fig. 3a).

STEP 4

Pull down on the throwline until the rope passes through the branch union and back down to where you can reach it. If you can't get the rope through the union, pull it back down and double check the knots. Or try tying the rope to the opposite end of the throwline. Just by reversing the direction the rope approaches the branch union can make all the difference. After removing the throwline, the rope is now ready for its intended purpose.

MUST KNOW KNOT: Pile Hitch

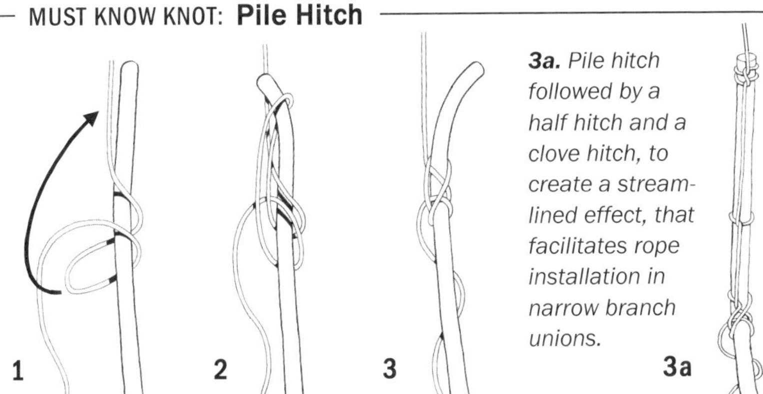

3a. *Pile hitch followed by a half hitch and a clove hitch, to create a streamlined effect, that facilitates rope installation in narrow branch unions.*

Untangling Tangled Throwline

GET USED TO IT: tangled throwline is a common occurrence when using this often irritating but indispensable piece of equipment. There's a lot you can do, however, to prevent it from happening in the first place and remedy it when it does.

Tips for Preventing Tangles

- Avoid throwing in areas of thick ground cover and tall grass.
- Clear the ground in the throwing area of twigs and debris that could potentially get caught in the line.
- Flake the line onto a tarp before throwing, or store it in a line bag, bucket, or cube for tangle-free deployment.

Buckets, bags, tarps, & cubes provide tangle-free deployment.

Tips for Removing Tangles

- If you have the persuasive ability, have a coworker untangle the line for you while you appear busy doing something "more urgent."
- If, however, you must perform this time consuming task yourself, begin by locating one end of the throwline and begin pulling it free from the tangles. Don't pull too tight or you'll turn the tangles into a tighter mess of knots. Make large loops in the line as you do this and the tangles will pull apart more easily or even come apart on their own.
- As you free up tangled line, flake it into a line bag, bucket, or cube. As you encounter a stubborn tangle or knot, you can feed the line container, and not the entire length of the line, through the enlarged loop.

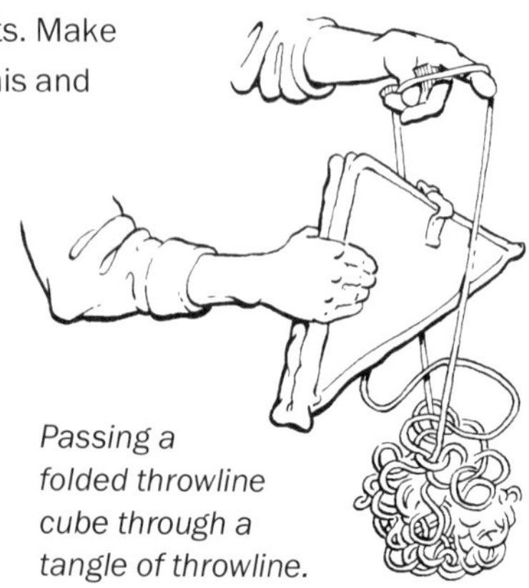

Passing a folded throwline cube through a tangle of throwline.

- If necessary, alternately work on both ends of the throwline, threading the shortest end through first.

SECURING A ROPE TO THE TREE

Once you have retrieved the end of the rope attached to the throwline (as described on page 103), it may be necessary to flip the line, or yank it, so it is properly positioned in the branch union and as close to the tree trunk as possible. This helps prevent the rope from breaking or sliding off an anchoring limb once tension is applied to the rope. When the rope being installed is a climbing line for a stationary rope system or to be used as a pull line for felling operations, there are two ways to secure the rope to the tree: with a canopy anchor or a basal anchor. Lines used for lowering limbs in natural-branch union rigging are not secured in either fashion as they are "moving lines" for hand held lowering or used with a lowering device.

Canopy Anchor

A canopy anchor is typically used with climbing lines and pull lines when they can be isolated around the main trunk or central branch union of the tree or a large branch in the canopy without any interfering branches. This arrangement offers the most secure tie-off, and uses the least amount of rope. The most simple way of securing the rope is to tie a ***running bowline*** (page 107) and pull the knot up to the anchor point in the tree. If any small branches interfere with the knot's ascent, give the rope a hard yank to break it free. Set the knot in place by giving it a firm tug.

The main drawback of this method is that the line is not retrievable. Of course, this does not matter in the case of a pull line, which can be retrieved after the tree is felled; likewise for climbers who plan to reposition the line after reaching the anchor point. However, with climbing techniques where it does matter, several specialized retrieval methods exist for this purpose, one of which is tying a retrieval line (such as a throwline in the inset above) to the bowline knot before running it up to the anchor point.

Basal Anchor

A basal anchor is used when the rope passes over interfering limbs, which prevent the use of a canopy anchor. This is often the case with conifers and thickly crowned deciduous trees. Here, the running bowline is tied around the base of the tree. In the case of a climbing line, there is also the option of securing it around the base of a neighboring tree, if the climber prefers. The best thing about basal anchors is that the rope does not have to be isolated in the canopy. This makes rope installation easier, faster, and an easy way to retrieve the rope from the ground. However, a basal anchor doubles the load on the primary support point and requires more rope length when being used as a pull line than does a canopy anchor.

The easiest way to anchor a climbing or pull line around the base of the tree is, once again, by using a running bowline (see opposite page). Make sure the bowline is secured around the trunk in such a way that it won't slide up the trunk once a load has been applied. When possible, secure the line below any trunk protrusions or swellings to prevent this from happening. In the case of a pull line, make sure the bowline is tied above the location where the notch and back cut will be made.

View from above of rope passing over multiple limbs in the canopy.

MUST KNOW KNOT: Bowline & Running Bowline

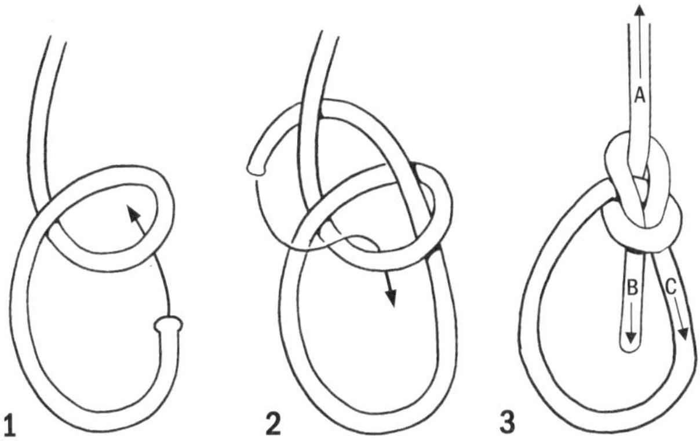

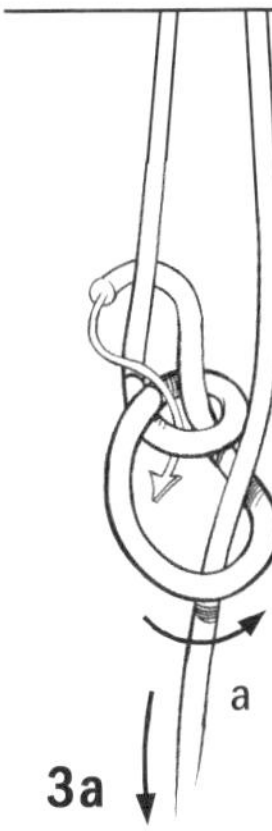

Common bowline

Running bowline

Tying the common bowline: *(1) "The rabbit (the working end of the rope) comes out of the hole" (the loop formed in the standing part of the rope)... (2) "goes around the tree" (the standing part of the rope)..."and then back into the hole." 3) Dress and set the knot by pulling on part A with one hand and parts B & C with the other.*

Tying the running bowline: *(3a) Tied exactly like the common bowline except the rabbit goes around standing part "a"* **before** *coming out of the hole. Pull down on standing part "a" to slide the knot up to the anchor point.*

When you learn the BOWLINE KNOT you can...

- *...use it to send up another rope to the climber (fig. 1).*
- *...use it to attach a throwline to a throw-bag (fig. 2).*
- *...more easily learn the double bowline, bowline on a bight (fig. 3), and running bowline (fig. 3a above).*

Figure 1

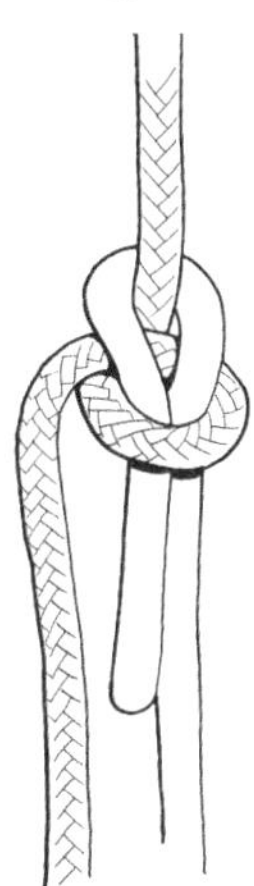

Figure 3

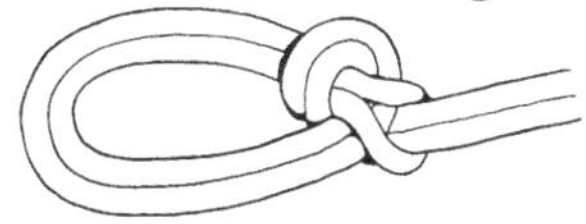

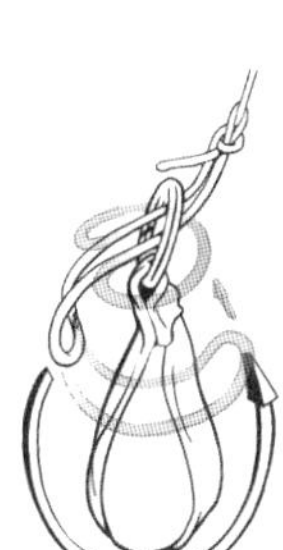

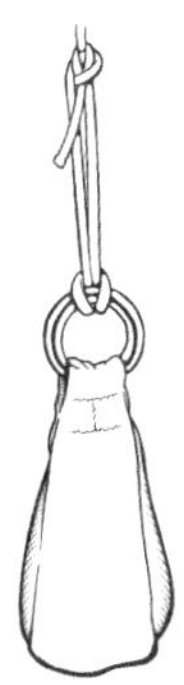

Figure 2

LOWERING LIMBS & TRUNK WOOD

During tree pruning and removal operations it is a common practice to lower limbs and trunk wood to avoid hazards and obstacles below. Often referred to as *roping down* or *rigging down* wood, this procedure is a team effort between the climber and ground worker (or workers). It is probably safe to say that there is no greater display of teamwork in all of tree work than when the two are working together to rig down wood. There probably is no aspect of tree work as dangerous either. These two facts are the reasons why it is so thrilling for others to watch.

The ground workers primary role in rigging operations is to run the ropes and maintain the drop zone. In its most basic sense, running the ropes means to hold on to and manage the rope securing the limb or tree section and lower it safely to the ground. This operation may or may not include the use of rigging equipment such as pulleys, blocks, or lowering devices, but it always involves ropes. The variety of rigging equipment and techniques available to the arborist is extensive and beyond the scope of this book to discuss in its entirety.[14] What follows is an introduction to this topic. The rest you will learn on-the-job or on-your-own by seeking other related resources, such as those listed on page 173.

Methods of Lowering

There is more than one good way to lower wood from the tree. But a skilled ground worker will choose the method that is the *best* one for the situation at hand. Some of the factors which help determine that choice include: (1) the size (length and weight) of the wood being lowered, (2) the amount of friction present at the rigging point, and (3) the availability of rigging equipment.

Hand-held Lowering

Small limbs and lightweight wood can often be lowered by hand—no rigging equipment required, just a good grip on the line, (while wearing gloves) and good judgment of the weight you can hold. If necessary, hold the rope against your hip to add friction and gain more control. ***Never wrap the rope around any part of your body!*** If, however, there is any doubt you can handle the load by hand—don't. Instead, add friction by taking trunk wraps or wraps on a lowering device.

Tips for Trunk Wrapping

- Whenever possible, take wraps around a neighboring tree. This will lessen the angle of the rigging line and the likelihood of the rope sliding up the trunk as wood is being lowered.
- It is much easier to carry and wrap the rigging line around the tree if it is stored inside a rope bag.
- Lower the wood slowly to reduce abrasion and heat friction on the rope, especially when wrapping around rough barked trees.

To add more friction and control, add more wraps around the tree.

Trunk Wraps

As the size and weight of the wood being removed gets heavier—beyond what can be safely managed by hand-held lowering—additional friction must be introduced into the rigging system to gain more control. This can be accomplished by simply wrapping the rigging line around the trunk of the tree being worked on or a neighboring tree. Be careful, however, when wrapping around a neighboring tree that the rope doesn't cause damage to the tree's bark. Before modern day lowering devices were introduced to the profession, this was the most common means of lowering wood.

Taking trunk wraps is still an effective technique when you understand its applications and limitations. To its advantage, trunk wrapping requires no additional equipment, is quick to employ, and is a relatively easy technique to learn. On the other hand, since trunk diameter and bark texture varies from tree to tree, it can be difficult to judge and control the amount of friction in the system. Trunk wrapping can also cause excessive wear on the rigging line, especially when wrapping around rough barked trees. But for occasional rigging applications with light to medium weight loads and when you need added friction in a hurry without having the time to set up a lowering device, trunk wrapping is often the best choice.

Lowering (Friction) Device

A lowering device (friction device) is a piece of equipment used to provide friction on a load line for controlled lowering of limbs and tree sections. There are a number of lowering device options available, such as a carabiner with Munter hitch, figure-8 device, Port-a-Wrap™, and bollard/winch systems. All of these are effective when used in the proper application, but the Port-a-Wrap tool (and others like it) are probably the most versatile of them all and therefore the one we will discuss more fully here.

With a Port-a-Wrap, users can easily add or subtract wraps—increasing or decreasing friction—which can be taken anywhere along the length of the rope without feeding the end through the device. The tube/post of the device is consistent in diameter and its surface smooth, thus minimizing rope damage and offering better and more predictable control than when tree/trunk wrapping. Finally, additional equipment can be incorporated with the Port-a-Wrap, such as fiddle blocks (mechanical advantage pulley system) so the lines can be tensioned and limbs lifted.

Typically, the Port-a-Wrap is secured to the base of a tree with a single-eye sling, also known as a dead-eye spliced eye sling, using either a ***timber hitch*** (page 112) or a cow hitch (page 172).

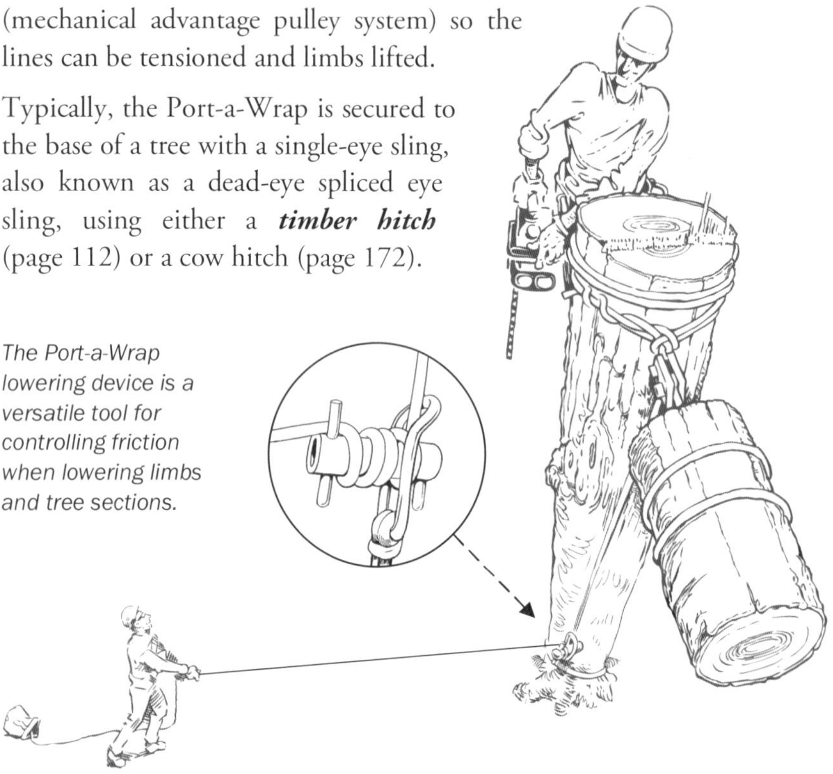

The Port-a-Wrap lowering device is a versatile tool for controlling friction when lowering limbs and tree sections.

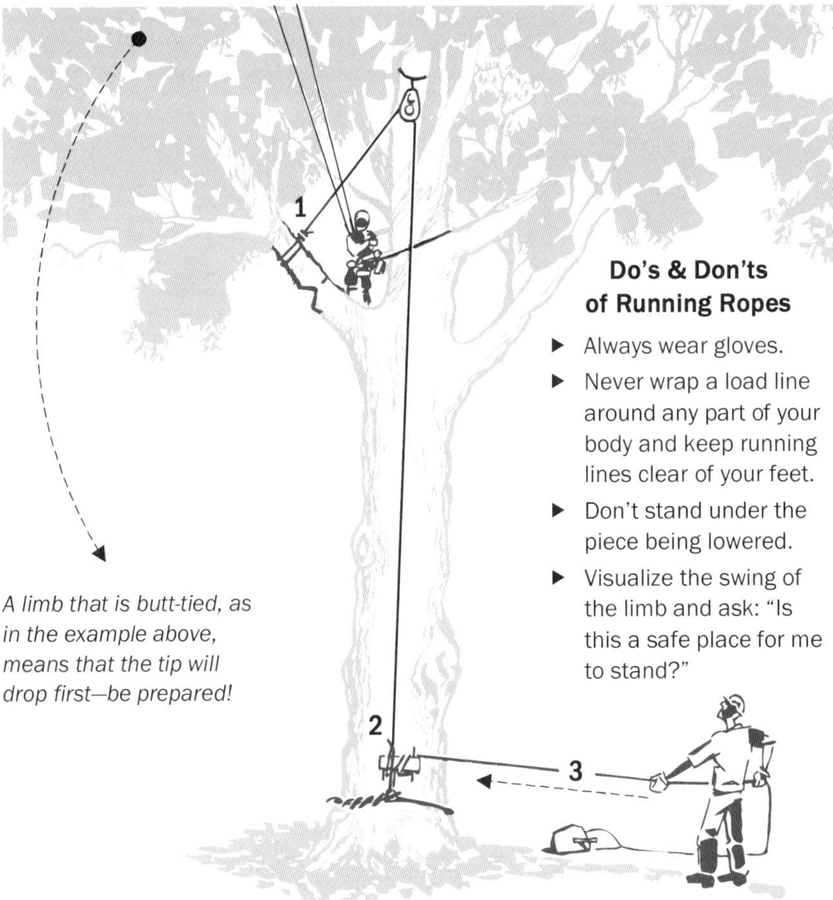

A limb that is butt-tied, as in the example above, means that the tip will drop first—be prepared!

Do's & Don'ts of Running Ropes

- Always wear gloves.
- Never wrap a load line around any part of your body and keep running lines clear of your feet.
- Don't stand under the piece being lowered.
- Visualize the swing of the limb and ask: "Is this a safe place for me to stand?"

Basic Limb Removal Procedure

1. The climber above has tied the load line to the butt end of the limb being removed. The climber waits until a "Ready here," or "On belay" command is given by the ground worker before cutting the limb.
2. The ground worker takes slack out of the line and adds enough wraps on the lowering device to lower the limb in a controlled manner. The heavier the limb, the more wraps—unless you are commanded to "let it run" (see page 113).
3. Once the limb has been cut and the swing controlled, the groundie can gradually release tension on the line and slowly lower the limb. In some instances it may be necessary for a coworker to help guide the limb away from hazards and obstacles such as power lines, roofs, gutters, and windows.
4. Once the limb is safely on the ground, the rope can be untied, sent back to the climber (see page 112), and the limb dragged out of the way. Repeat.

MUST KNOW KNOT: Timber Hitch

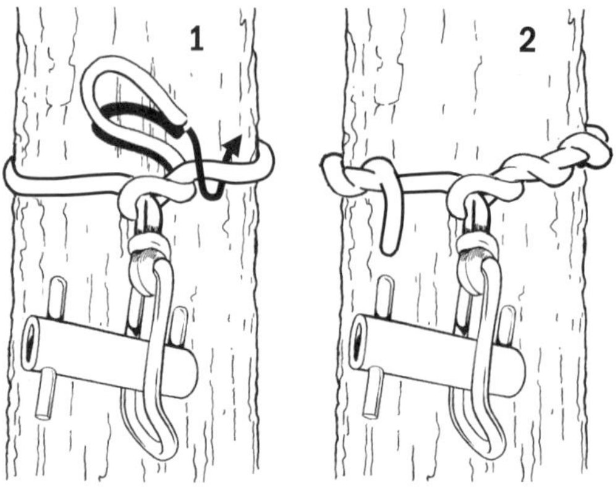

How to tie: *(1) Pass the rope of the sling around the tree and make a turn on the standing part (reversing direction) and make the first tuck back under itself. (2) Continue making a series of tucks that spiral back under itself. Spread the tucks out evenly around the trunk at least five times.*

Returning the Rope

Once a limb has been lowered to the ground, return the rope to the climber as soon as possible. Before you do, tie a slip knot (page 96) in the end of the line. This will prevent the rope from accidentally running through the pulley or block.

First estimate the distance from the climber to the pulley and then tie the slip knot in the line at this point. This will position the working end of the load line close to the climber's position in the tree, where the climber can regain use of it after pulling the slip knot apart. Make sure to tie the slip knot tightly so the weight of the rope below the knot doesn't pull it out, but not so tight the climber can't pull the knot apart either. Also, remember to alert the climber with an "up rope" command before sending it back up.

Additional Rigging Terms & Concepts

Aside from the all important "Stand clear" command, and the "All clear" response (page 97), there are a few other important commands, terms, and concepts to learn before you begin lowering wood from the tree.

Shock-load—A sudden force placed on a rope, rigging tool (blocks, pulleys, and slings), tree, or other object when a moving load comes to a stop.

"Hold fast"—This command, along with "hold it tight," or others like it, means that the ground worker is to hold tight on the line, without letting any of it pass through their hands or lowering device, to prevent the load from dropping after it is cut. For this to happen, enough wraps will need to be taken around the tree trunk or lowering device. Failure to hold fast the load could cause the piece to strike a hazard or obstacle below it.

If you're not sure how many wraps you should take, take more than necessary. You can always take off extra wraps even as you are lowering, but it is difficult to add more. Always ask the climber whether you are to hold fast to the line or "let it run" (see below) if you are uncertain. To mistakenly hold fast to the line when the climber wants the load to "run" can have disastrous consequences.

"Let it run"—In contrast to the hold fast command is the "let it run" command. This means that the ground worker is to *gradually* increase tension on the load line to slow down the speed and force of the piece being lowered before bringing it to a full stop. This is a tricky skill to master. Not enough wraps on the tree or lowering device and the piece will fall too fast and could potentially strike a hazard or an obstacle below; if too many wraps are taken, the piece will come to a sudden stop and shock-load the rope and rigging, which could cause either to fail.

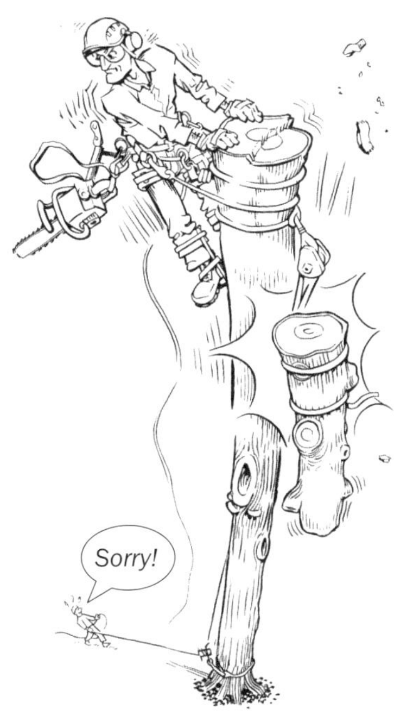

Likewise, the tree itself could even fail and fall if it has serious structural defects. At the very least, the effect of a "no-run" can cause the climber to be bounced around on the trunk and possibly injured. It is no wonder then that mastering this skill is considered the height of achievement in ground work operations. Get it right and the climber will be singing your praises; screw it up and you'll have some apologizing to do. Practice this skill in non-critical situations.

Rigging point—A rigging point is a place in the canopy or on the trunk of the tree that the load line passes over, such as a branch union or the sheave of a pulley or block, to control limb and tree section removal in rigging operations. The rigging point may be above the piece being lowered or below, as shown in the illustration at right.

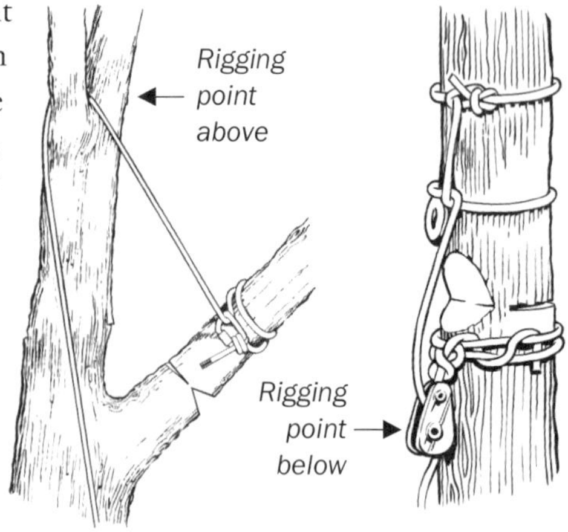

Of the two, having the rigging point positioned above the piece places less load on the rigging and offers better control when lowering. But when it's time to rig down the top of the tree and trunk wood, the rigging point must be positioned below the cut (unless the rigging is set in another tree). In this instance, it's crucial that the groundie let the line run smoothly.

Sweating the line—To "sweat up" the line (also known as "swigging") is to pull sideways, rather than straight down, on an already taut lowering line, in order to take as much slack out of it as possible. In most instances, one person can effectively sweat the line to take up more slack. But in more critical rigging situations, it will require two people to pull sufficient slack out of the line. As one person pulls at a right angle to the rope's length, the other quickly takes up the slack on the lowering device. It may take several attempts of sweating the line to take up all the slack that you potentially can.

Friction—One of the most simple and understandable definitions of friction comes from a children's book, *The How and Why Wonder Book of Machines.*[15] It defines friction as "the resistance that is caused when one object moves against another." As it relates to lowering wood, friction occurs in a number of ways. First, friction occurs at the rigging point when a line moves across a pulley sheave or a branch union. Secondly, friction occurs on the ground when rope wraps are taken on the trunk of a tree or on a lowering device. But regardless of how and where friction occurs, it is primarily the ground workers responsibility to manage it properly.

Generally speaking, the more friction that occurs at the rigging point in the tree, the less friction (or wraps) needed on a lowering device on the ground. Again, determining how many wraps (if any) are necessary, depends on all sorts of factors. This is something that can be determined only through experience and good communication between the climber and ground worker.

Maintaining the Drop Zone

The drop zone or landing zone, is the area beneath the tree where cut branches or wood sections will be dropped or lowered. The possibility of being struck by an object from above is never greater than when working in this area, which is why it is sometimes referred to as the kill zone or death zone. The primary goal in maintaining the drop zone is to keep the area free of anything or anybody that will hinder safe work operations. For the most part, this will involve keeping ropes from getting tangled and damaged, moving limbs, logs, debris, and equipment out of the way, but also such things as preventing bystanders or pets from entering the area.

Rope management—It is quite common to use several lines, such as climbing, load, and tag lines (see glossary p. 174), at the same time during rigging operations. Good rope management involves keeping them separated and from getting tangled with limbs on the ground. Believe it or not, every year climbers are pulled out of trees when ropes are tangled in brush piles and mistakenly fed into chippers. If a limb does become tangled with a line, free it up immediately. Keeping ropes tangle free and litter free, will be an ongoing task for the ground worker and one of utmost importance.

Good rope management also means: ***don't step on the ropes!*** And don't allow others to do it either. This can be a real challenge when ropes are laying all over the place in the tight confines of the drop zone. One way to avoid this problem is to keep the ropes close to the base of the tree. Also, use rope bags, buckets, or barrels. They are a great way to keep rigging and climbing lines clean and organized, and make it easy to move ropes out of the way when necessary. Once a line has been detached from a limb, send it back to the climber. If a pulley or block is used at the rigging point, return the line using the technique on page 112. Depending on where the climber is positioned in the tree, it may be necessary to "whip the line" so he or she may grab hold of it.

Branch management—Moving limbs, logs, and other tree debris out of the drop zone, or at least from underfoot, is probably the biggest challenge in maintaining the drop zone. Things on the ground can get crowded and chaotic quickly when the ground worker can't keep up, which makes work less productive and safe.

Staging limbs outside the drop zone.

The ideal, but rarely practical, strategy is to drag limbs to the chipper as they are being dropped or lowered from the tree, where they can be "staged" (neatly piled) to be chipped later. More likely though, the ground worker will only have time to drag or toss limbs as far from the work area as is necessary to prevent rope tangles and trip hazards. Material should always be staged with the butt ends facing the drag path. Larger wood is cut up smaller and moved out of the way to be dealt with later.

When the debris on the ground poses a safety hazard, it's time to let the climber know that you need time to get caught up. Rather than being an inconvenience, most climbers will appreciate the break, provided of course, it's not too long a wait. During removal operations, especially ones requiring lots of rigging, it is best to have two groundies working together: one, or both, to run the ropes, the other to cut and drag limbs to the chipper, or, at the very least, out of the drop zone.

Time management—The key to efficiency is using your time wisely. When, for instance, the climber is busy doing something that doesn't require your help—say, moving to another location in the tree—use the time to get caught up by dragging or staging brush, raking up and putting small debris ("ankle twisters") in a nearby wheelbarrow, moving unneeded tools out from the work area, or stuffing a line in a rope bag. Or you might use the time to take a swig of water as you watch the climber and think what he or she might need help with next. Finally, get in the habit of asking yourself, "How can I make what's going on here more efficient?"

Attention management—Finally, maintaining the drop zone means managing your attention—something we've repeatedly emphasized in this book and will do so again here—it's that important. To manage your attention is to be present in the moment, to "be here now," control distractions, and maximize your focus on the work at hand, but not so focused on any one thing that you are unaware of other important things. There is a lot going on at the job site, and to effectively take it all in you need to "keep your head on a swivel," as some have described it. Failure to do so can be deadly, as arborist Donald Blair so graphically reminds us:

> "If you close off your awareness to what is above, below, behind, left or right, you have effectively stuck your head in a box. Get your head out of the box or you may end up with your whole body in one—a pine box."[16]

Ground workers need to be constantly looking up—One of the most frequent tips climbers have for groundies is, "pay attention to the climber," "keep looking up," or similar phrases. This will help keep you safe, of course, but it's essential to know what's going on above in the tree so you can be more intentional with your efforts on the ground. Be alert for falling objects for sure, but also for canopy hazards the climber might overlook. Anticipate the climber's needs (page 99) and watch for a break in the action so you can get caught up with other chores—or take a much needed short break.

"Keep your head on a swivel."

Ground workers need to be constantly looking around—One good reason to be constantly looking around while working in and near the drop zone is to keep out the curious customer, the wandering walker, and the unleashed pet. They often appear out of nowhere and when you least expect them. They are as much a hazard and nuisance as a nearby hornets nest. Like hornets, you need to make them go away ASAP!—only do so in a friendly fashion. Taking frequent glances of your surroundings will also keep you informed of the whereabouts and activities of the rest of the crew. Then, of course, there are the limbs and lines that need your constant attention.

PAY ATTENTION!—To Be Learned (lur'-nid)

NO DOUBT, most of what you learn about tree work will come from hands-on-instruction, and related books, articles, and training videos. But you can learn just as much by paying attention to your teachers at the school of hard knocks. That's the school where the lessons learned come the hard way—from mistakes and failures.

Learn from your mistakes **—** The lessons learned from making mistakes are not quickly forgotten, which is probably why they make such good teachers. Many mistakes, especially those of the "near-death" variety, stick with us for a lifetime. But even the less serious types of mistakes are often memorable, especially when they are embarrassing or painful.

When you make mistakes, learn from them. Whether it is a colossal failure or a minor mistake, ask yourself—or another more experienced worker—"What did I do wrong, what were the consequences, and what should I do different next time?" The next step is to do something about it. Rather than grumble or make excuses about a particular failure, like how you forgot, once again, to properly tie a quick hitch or bowline, take home a hank of rope and practice the darn thing!

Learn from the mistakes of others **—** The lessons learned from your own mistakes may have a lasting impact, but they come at a cost. You may have learned to wear gloves when running the ropes, but you paid for the lesson with blistered hands. That alone is good enough reason to learn from other's mistakes. Take the advice of Alaskan bush pilot Mort Mason: "Learn from the mistakes of others. You won't live long enough to make them all yourself"—nor would you want to.

Aside from learning from our coworkers, we can also learn from the experiences of those we don't even work with or know. For example, each month the *Tree Care Industry Magazine* publishes accounts of reported accidents ("incidents", if you prefer) related to tree work.[17] These briefs are intended to inform, as well as shock us to attention. With titles like, "Worker Killed by Chain Saw Cuts," "Man Loses Arm in Chipper Incident," "Ground Worker Killed in Struck-by," we would be fools not to listen and learn from these tragic mishaps. As for me and my crew, we make it a regular practice to read aloud these accident briefs and discuss how they might affect us.

CHAIN SAW SAFETY & OPERATION

Up to this point in the book (and on the job) it is quite possible to have gotten by without even using a chain saw, other than to fuel or send one up to the climber. That, however, is about to change. In chapter 2 you learned how to maintain a chain saw and what personal protection equipment you need to wear when running one. Now, over the next seven pages, you will learn the basics of chain saw safety so that you can use this tool to cut up limbs, logs, and downed trees. And eventually, to fell trees as well. If you don't want to become one of the estimated 36,000 people who are treated in emergency rooms for chain saw injuries each year, you will devote considerable time and effort mastering this tool.

Starting a Chain Saw

Avoiding chain saw injuries begins by learning how to properly start your saw using either the *ground start* or the *leg lock* method as follows:

Ground Start Method

1. Apply the saw's chain brake and place the saw on level ground clear of rocks or debris.
2. Position the on-off, choke, and throttle switch according to instructions in the owner's manual.
3. Grip the front handle bar, keeping your left arm straight and elbow locked as you apply a slight downward pressure. Place the toe of your right foot in the rear saw handle.
4. Pull the starter cord slowly, using the right hand until you feel it engage. Then give quick strong pulls until the saw "coughs," at which point you can turn the choke off and pull again until the saw starts.

Ground starting offers the most stability.

5. Pick up the saw, gently squeeze the throttle trigger to simmer down the saw's engine RPMs (revolutions per minute), release the chain brake, and begin cutting once the saw is warmed up.

Leg Lock Method

This is the preferred method of most ground workers and the only one used by climbers when starting a chain saw aloft.

1. Apply the saw's chain brake.
2. Position the on-off, choke, and throttle switch according to instructions in the owner's manual.
3. Grip the front handle bar firmly with the left hand and clamp the back of your right thigh over the back handle of the chain saw.
4. Pull the starter cord slowly, using the right hand until you feel it engage, then follow the rest of the starting procedure as described by numbers four and five on the previous page.

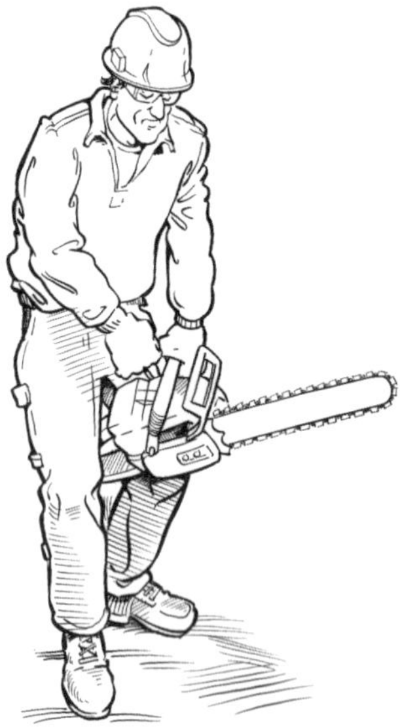

Leg lock starting offers the most convenience.

Drop Starting

This is a starting method where the chain saw is pushed away from the body while simultaneously pulling the start cord. It is most frequently performed by inexperienced tree workers and the veterans who simply refuse to learn a proper and safer method.

For the record, drop starting is a dangerous and unacceptable method of starting a chain saw. What's more, drop starting is *prohibited* by ANSI Z133. If the saw tip were to strike a hard object, with the chain brake off, the saw could *kick back* (see page 123) toward the operator, as well as dull the saw chain. Drop starting also causes excessive wear and tear on the starter cord.

Proper Grip

Chain saws are designed for right-hand use, that is, the right hand grips the rear handle, operating the throttle, and the left hand grips the forward handle—no exceptions. Keep both hands on the saw anytime it is running. The only safe way to grip the front handle is with the thumb wrapped *under* the handle. Never grip the handle with the thumb on top or off to the side!

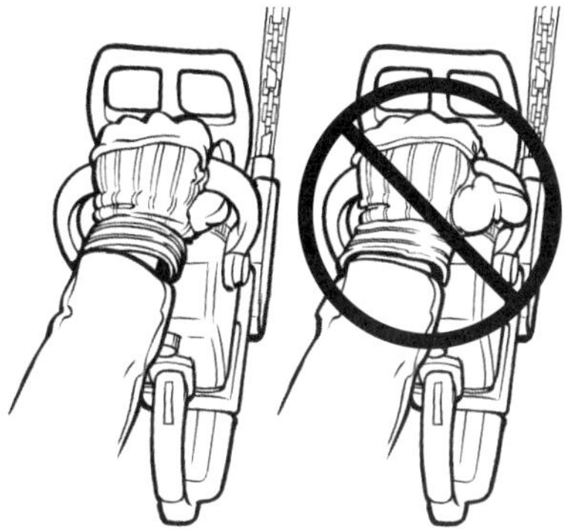

Correct Incorrect

Proper Cutting Stance

Maintaining a proper stance while operating a chain saw plays an important role in personal safety and work efficiency.

Stand left of the path of kickback.

- Keep your left arm straight and maintain a firm grip on the saw with both hands on the handles. Wrap fingers and thumbs completely around the handles.
- Position yourself slightly to the left of the saw while cutting. If kickback occurs, the saw will likely pass away from your body.
- Bend your knees slightly and keep the saw close to the body to increase control and reduce fatigue.
- Maintain sure footing and good balance when cutting. Move any ground debris and obstructions that could hinder your balance and movement.
- If the saw is properly sharpened you should not have to push down with excessive force on the saw to cut. Let the weight of the saw provide most of the cutting pressure.

Carrying the Chain Saw

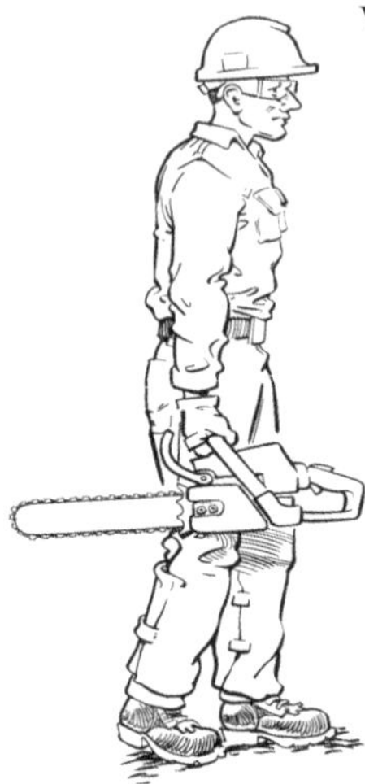

When you carry the chain saw to and from the cutting site, hold the saw at your side with the bar pointing backward away from the direction you are walking. Carrying the saw in this position will prevent the bar from catching on brush and, should you fall, will minimize your risk of landing on the bar and chain. Another acceptable way of holding and carrying a saw is like that shown on the cover of this book.

If you carry the chain saw more than two steps or take one hand off the running saw, the chain brake must be engaged. Of course, you should always make sure the saw is shut off whenever you leave it unattended.

Understanding Chain Saw Reactive Forces

Another critical safety concern for preventing chain saw injury is understanding the reactive forces that occur when the saw bar and chain make contact with wood or another object. There are three main reactive forces that can potentially occur at any given moment while cutting wood—push force, pull force, and kickback. To be unaware of the constant presence of these forces is a surefire recipe for disaster, especially in the case of kickback, whose recipients receive injuries that are often severe or fatal.

Push force—Whenever wood is being cut with the top of the bar, as when undercutting a log in bucking operations, the forward motion of the chain will tend to push the saw away from the wood and toward the operator (fig. A next page). This can be a dangerous situation if you encounter such a force while using poor cutting stance, gripping the saw improperly, or are physically and mentally unprepared to respond to its effects, which can be powerful when operating a larger saw.

Pull force—Whenever you cut wood with the bottom of the bar, which is most of the time, the cutter teeth making contact with the wood tend to pull the chain saw (and you) toward the wood (fig. B next page). This can be especially hazardous when cutting smaller pieces of wood, which can be thrown forcefully toward the crotch area (men, you've been warned).

Figure A: Push force

Figure B: Pull force

The effect of pull forces can be minimized by keeping the saw head close to the wood being cut and engine and chain speed relatively high. Be prepared for reactive forces by maintaining a good cutting stance (page 121) and keeping a firm grip on the saw. A properly sharpened saw will also help lessen the effects of pull force. Chains in which the depth gauges have been filed down too much tend to accentuate pull forces dangerously.

Kickback—There are two forms of kickback: rotational and linear. Rotational kickback occurs when the upper corner of the bar (kickback corner or "no zone") accidentally makes contact with an object (shown below). This contact forces the chain to a sudden stop and causes the saw to push itself off the object with a violent rebound toward the operator. Linear kickback is a more violent form of the push force previously described that occurs when the wood closes, pinches the saw chain in the cut, and pushes the saw toward the operator.

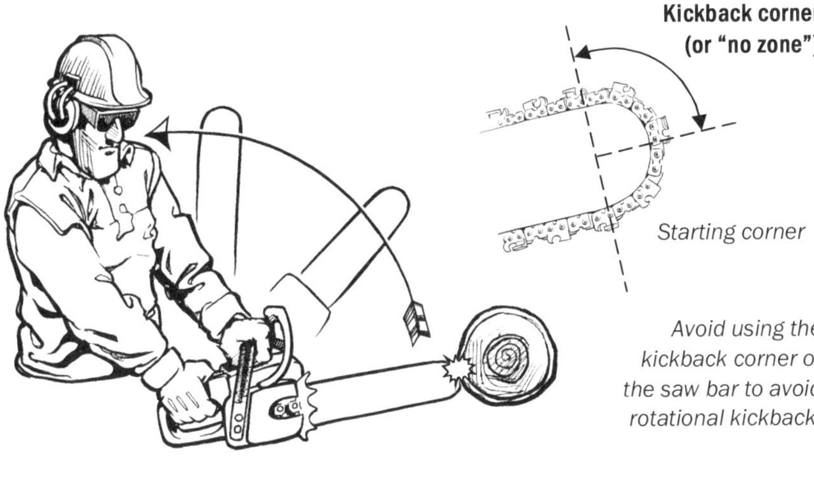

Avoid using the kickback corner of the saw bar to avoid rotational kickback.

In both cases (but especially with rotational kickback) there is no warning and little time to react when it occurs. From start to finish, the whole event takes place in a fraction of a second—you'll never know what hit ya.

From start to finish, the whole event takes place in a fraction of a second—you'll never know what hit ya.

Rotational kickback occurs most commonly while bucking logs and the bar tip makes contact with other logs piled behind the ones being cut or when using the tip of the bar to remove limbs close to the trunk. Wearing personal protective equipment will certainly help prevent or minimize injury if kickback does occur, but precautions, like those described on the next page, must be taken to help prevent injuries from happening in the first place.

Precautions for Helping Prevent Kickback

- Avoid use of the upper quadrant of the bar nose, or kickback corner.
- Clear the cutting area of any underbrush, branches, logs, rocks, or other solid objects that could come in contact with the bar nose.
- Hold the saw firmly with two hands with the left thumb wrapped *under* the front handle.
- Maintain a stable body position before beginning a cut.
- Know where the tip of the saw bar is at all times during operation.
- Keep the bar tip clear of any obstacles.
- Use saws equipped with a functioning chain brake.
- Keep your chain saw properly sharpened.
- Maintain a high saw speed when entering or leaving a cut.
- Use a low-kickback saw chain whenever possible.

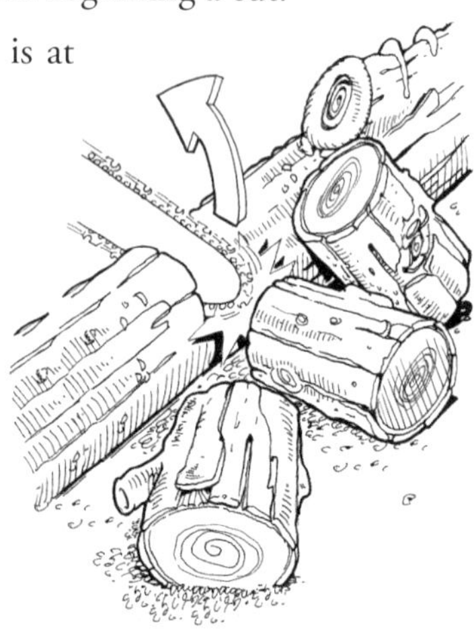

Avoid kickback! Keep the bar tip clear of obstacles.

Six More Ways to be Chain Saw Safe

Aside from what's already been said about chain saw safety in this chapter, including the necessity of wearing PPE presented in chapter 2, here are six more ways to be chain saw safe.

1. ***Use only saws equipped*** with the proper bar length, front hand guard, anti-vibration features, chain catcher, spark arrester, and a functioning throttle trigger interlock, chain brake, and stop switch.
2. ***Keep your saw in tip-top shape*** by learning how to inspect, clean, and maintain it on a regular basis as described on pages 32-47.
3. ***Maintain a distance of at least ten feet*** when working near another chain saw operator (see page 141).
4. ***Stay off ladders and out of trees*** while operating a chain saw unless you have been specifically trained to do so.
5. ***Never cut above shoulder height:*** thus sayeth ANSI Z133-2017: 6.3.9.
6. ***Never operate a chain saw*** while under the influence of alcohol (or even a hangover) or drugs. Even some cold medications can adversely affect your ability to work safely.

On Buying a Chain Saw

"A CHAIN SAW is not something to be bought at the garden center on a Saturday morning, with your children's fingers sticky from ice cream, your wife impatient to get on with the rest of her day, and the parking meter about to expire. Like a man's choice of hunting rifle, car, and sound system, the selection of a chain saw is something to linger over. There are catalogs to be studied, specifications to be compared, more catalogs to be plowed through in peace and quiet. Every decimal point relating to the horsepower and the vibration level of the handle must be minutely studied before the final choice is made. Only then has the groundwork been laid for the development of a genuine relationship with your chain saw."

—Lars Mytting, from *Norwegian Wood*

** For some of you reading this, it might be helpful to know that my wife owns several guns, is an avid hunter, picks out the family car, and has her own prized chain saw that no one dares touch.*

BASIC TREE FELLING TECHNIQUE

There is much more to felling a tree than making a few cuts with a chain saw and yelling "Timber!" as you run for your life and hope for the best. Tree felling is an art and science that requires thoughtful planning and skillful cutting to ensure the tree lands where it is supposed to and without anyone getting hurt (or killed) or anything getting damaged. This is a calculated accomplishment, not an exercise in luck, though, by the grace of God, many do get "lucky" and live to tell about it. That is why tree felling is a skill typically not taught to beginning ground workers. But, when the time to learn does come, the basic principles that follow will provide you with a good introduction to felling trees safely.

Keep in mind, though, that entire books have been written about this topic which is presented here in only six pages.[18] Nevertheless, this material, coupled with an experienced and competent trainer, will be enough to get you started. At the very least it will help you know what's going on when you're watching others fell trees. You can begin by practicing on tall tree stumps or upended logs. As you become more competent in making felling cuts, you can then move on to the real thing and fell "easy" trees in non-critical situations, under the supervision of a competent instructor.

Assessing the Tree & Felling Site

Before felling a tree you must first inspect the felling site and the tree being felled. In the felling site inspection, you're looking for hazards and obstacles to avoid, something you already did earlier in the chapter. In the tree inspection, you are assessing the tree itself to determine four key things: (1) the tree's height, (2) the tree's lean, (3) the direction you wish the tree to land (the lay), and (4) your escape (retreat) route from the work area.

Determining tree height is explained on page 86. Determining the tree's lean is usually performed at the same time you estimate tree height. You'll need to know not only which direction the tree leans (by taking readings from two locations, 90° to each other), but also how much. Both factors will determine whether felling aids, such as felling wedges or a pull line, might be necessary. Determining the lay of the tree depends on the tree's lean and the presence of site hazards and obstacles. Finally, an escape route from the work area must be identified and cleared.

Making the Felling Cuts

All your efforts of preparation and safeguarding against accidents can be undone in a matter of seconds if you fail to make the felling cuts correctly. Therefore, it's important that you first understand how the *notch, back cut,* and *hinge*—the felling cut trio—work in unison to guide the tree into the intended lay. So first the definitions, then the mechanics of proper cutting.

The notch, also referred to as a face cut or face notch, is a wedge-shaped opening made by making two cuts on the side of the tree facing the lay. This opening allows the hinge to work and enables the tree to fall freely in the direction it was made. The size and depth of the notch opening will affect *how* the tree will fall, while the position of the notch on the trunk will determine *where* the tree will fall.

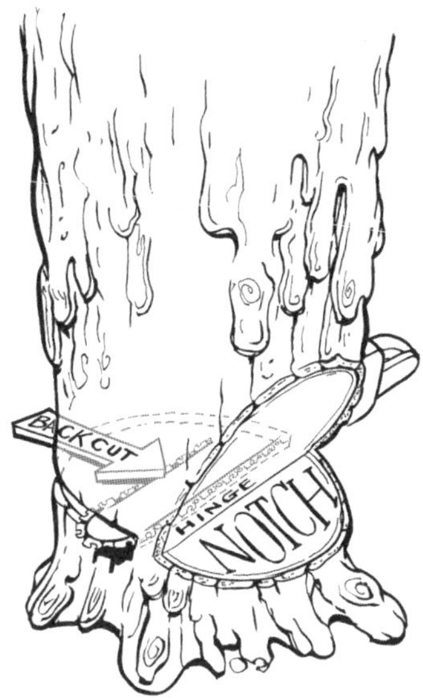

The "felling cut trio"—the notch, back cut, and hinge—work in unison to guide the tree into the intended lay.

The back cut is a single horizontal cut made on the opposite side of the tree from which the notch is made. There are instances when the back cut is made in several stages such as when felling trees with a greater diameter than the saw bar. The back cut removes a predetermined amount of wood fiber thereby establishing the thickness (width) of the hinge.

The hinge is the strip of wood left uncut between the notch and back cut. The primary function of the hinge is to steer and control the direction the tree will fall and prevent it from twisting or jumping off the stump. The length and thickness of the hinge are determined by several factors, such as how deep the notch is made, how much wood is removed with the back cut, and the tree's trunk diameter. The longer the tree stays attached to the hinge, the greater the degree of support and control you will have in guiding the tree into the lay.

Step 1: Cutting the Notch

The notch is always made before the back cut. A well prepared notch is essential for the proper functioning of the hinge in guiding the tree to the ground. There are four factors to consider when cutting the notch: notch placement, notch cutting sequence, notch size, and notch depth.

Notch placement—The first step in cutting the notch is to position it on the trunk in the exact direction you want the tree to fell. This may seem obvious, but even a slight error in notch placement can have a dramatic effect on where the tree will land. For instance, a notch cut only 10 degrees off from the intended direction will cause a 60-foot tall tree to land 12 feet off target.

Aiming the saw

The easiest way to position the notch is by using the felling, or *gunning sights*, marked on top of the shroud and housing of your chain saw. These sights can be a painted raised line or decal. They are arranged so they run perpendicular to the bar of the saw and the notch you are about to make. "Aim" the saw by sighting along this line directly toward the intended lay—the direction you want the tree to land. This is the location on the trunk where your first notch cut will be made.

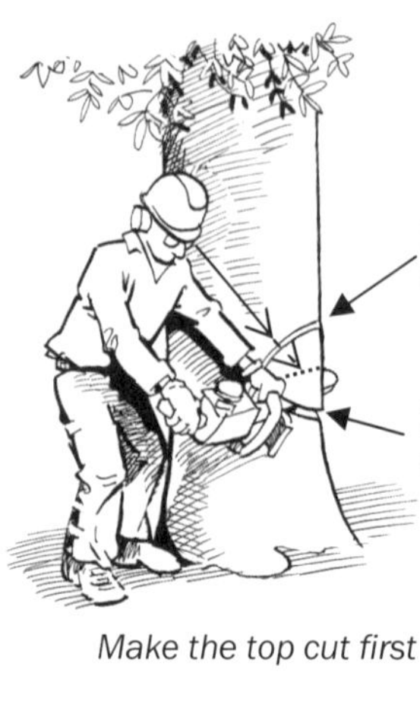

Make the top cut first

Notch cutting sequence—You should always make the top cut of the notch first and the bottom cut second. The saw kerf created by the top cut functions as a "window" you can sight down to watch for the emerging chain teeth when the bottom cut is made. As soon as they come into view, and the bottom cut meets the top cut, it is very important that you stop cutting so that the notch cuts meet evenly.

Notch size—The size of the notch opening is determined by the notching method chosen. The *common*, or *traditional notch* has an opening of 45 degrees. The increasingly popular *open-faced notch*, and the one presented in this book, has an opening of 70-90 degrees. The reason the open-faced notch has become so popular with arborists is because of what happens as the tree falls and the notch closes.

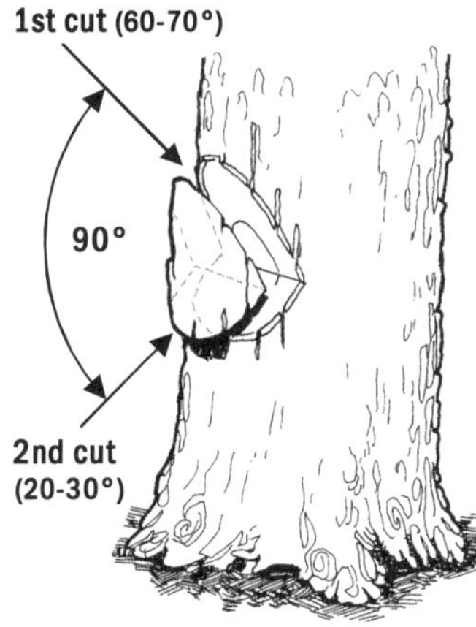

Make the top cut at a 60-70° angle to a point 1/5-1/4 the diameter of the tree. Make the bottom cut at a 20-30° angle to meet the first cut.

The common notch, with its 45-degree opening, closes when the tree has only fallen about half way to the ground. The resulting resistance from the closed notch causes the wood fibers of the hinge to break. The tree then separates from the stump before it has reached the ground—essentially falling out of control. The open-face notch, with its larger notch opening (70-90 degrees), allows the tree to remain attached to the hinge longer, closing when the tree has reached, or almost reached the ground. Consequently, the tree remains under control the entire time it falls.

Notch depth—Cutting the notch to a depth that is 20-25% (1/5-1/4) of the diameter of the tree is the current guideline among arborists. You can estimate this quickly from visual inspection alone. For example, if the tree is 15 inches in diameter, the notch depth should be about 3 inches using the 20% rule (shown below), slightly deeper with a 25% notch depth. Compared to the traditional 1/3 notch depth, the 20-25% rule performs significantly better in a wider variety of cutting situations. Besides, you can always cut a deeper notch if necessary, but you can never cut less.

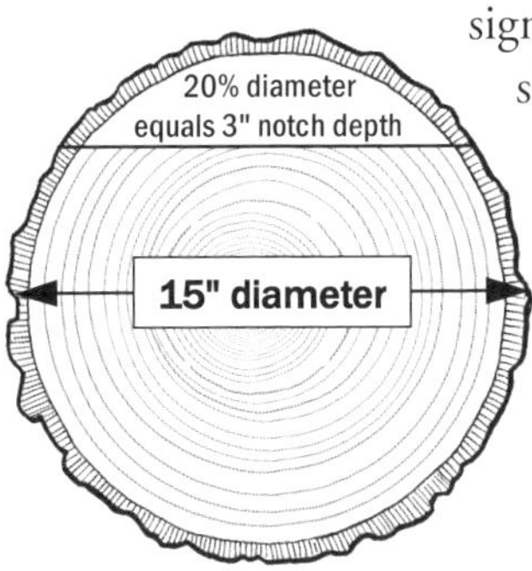

The depth of the notch also determines hinge length (the back cut determines hinge thickness). The ideal hinge length is 80% of the tree's diameter, which is usually what you end up with when using the 20-25% rule.

Step 2: Making the Back Cut

Unlike the notch, which allows you to make corrections for cutting errors, the back cut is unforgiving. You can't undo the removal of wood you've already cut.

Begin making the back cut by positioning yourself on the "good" side of the tree—that is, the side opposite a side lean, if it has one. If the tree has no significant lean, most cutters prefer to make the back cut from a position where their left shoulder faces the lay, as in the illustration above. This position feels most natural, comfortable, and offers the best cutting leverage since it utilizes the bottom of the guide bar.

Before making the back cut, give a clear and loud warning of "stand clear!" to all workers within the work area and any bystanders in the vicinity. Listen for a response of "all clear" or look for a visual signal, such as a "thumbs up," before proceeding. This same signal or one like it ("timber!"), should be given again as you are actually making the back cut.

When using the open-face method, make the back cut even with the apex of the notch, as opposed to 1-2 inches above as with the common notch. Also, cut inward evenly so the finished back cut runs parallel to the back of the notch. The goal here is to create a hinge that is of equal thickness throughout its length. How thick or wide you leave the hinge can be determined by a simple rule of thumb: tree diameter (at the point where the back cut is made) X 10% = the minimum hinge width (see example on next page). Hinge thickness should be determined before you make the back cut, and marked if necessary, so that you don't accidentally cut too far. The last thing you need is a hinge that is too narrow or cut completely through, and a tree falling dangerously out of control.

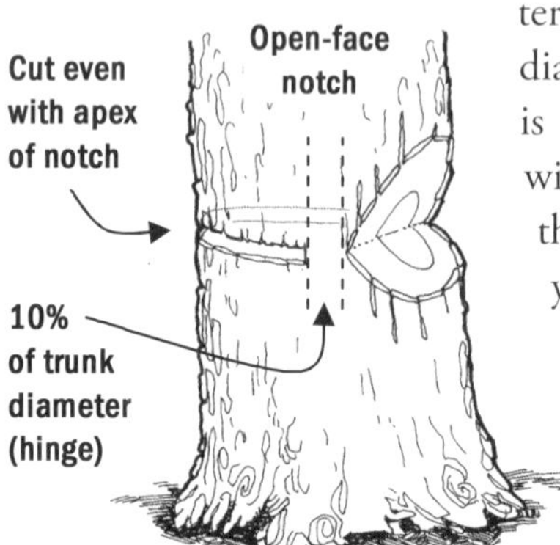

Know the position of the saw bar at all times and how much wood is being cut on *both* sides of the back cut. It is easy to focus your cutting attention on only one side of the tree and overlook what the bar is doing on the other side. As the chain teeth begin to approach the designated hinge area, cut slower to avoid cutting too much hinge wood. Remember, you can always cut more wood if necessary.

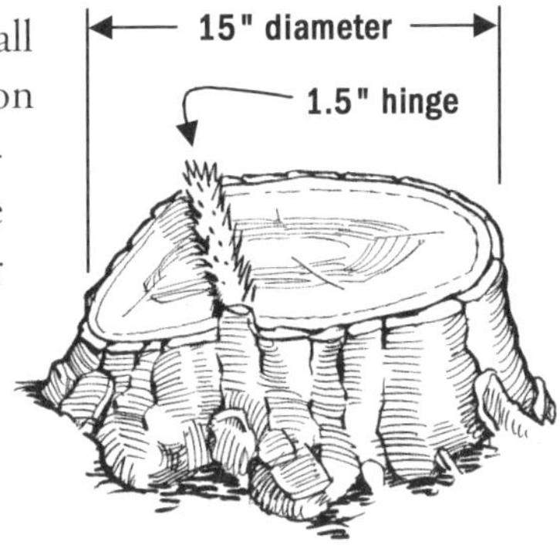

15" diameter X 10%=1.5"

Step 3: Retreating from the Work Area

When you have nearly finished the back cut and are approaching the predetermined hinge thickness, look up at the top of the tree for the first signs of movement. Only when the canopy is viewed against the sky or in relation to neighboring trees will the early signs of movement be evident, indicating the tree is beginning to fall. When this happens, begin retreating along your pre-determined path of escape for at least 20 feet from the tree opposite the lay at a 45-degree angle. Most felling injuries and deaths occur within a fifteen-foot radius of the tree.

Make sure you take the chain saw with you (unless the tree begins to fall unexpectedly or the bar is pinched) after shutting off the engine or activating the chain brake. As you retreat, watch the tree, as well as where you are going, to make sure it's falling in the right direction, or at least not toward you. This is also a critical time to watch for falling debris. Once the tree is on the ground, look up into the canopy of surrounding trees for any broken or hanging branches that could be hazardous to work under.

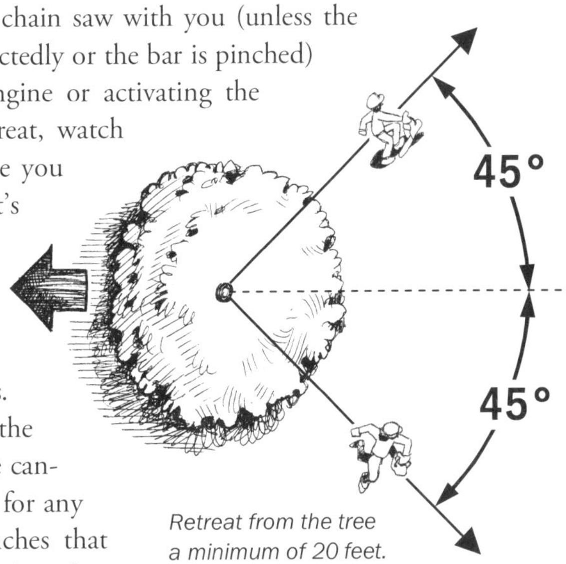

Retreat from the tree a minimum of 20 feet.

Felling a Set Back Tree & Freeing a Stuck Saw

Felling trees is not always as easy as step one, two, three. Things go wrong, such as when you're making the back cut and the tree sets back on the saw bar. This usually results from misjudging the trees lean and/or when an unfavorable wind picks up. Then, without warning, you have a stuck saw. This is a potentially dangerous situation that requires immediate action since the tree could fall anytime and in any direction. For this reason, never turn your back on a set back tree and get anyone not involved with the felling operation outside the felling radius of the tree. What follows are six options for rescuing a stuck chain saw and getting the tree to fall into the lay. Whenever possible, have someone ready to remove the saw at the first opportunity. Last-second attempts to save a saw as the tree falls are dangerous and not worth the risk. Consider removing the power head from the bar to prevent damage to the saw should the tree fall unexpectedly.

Option 1: Get Help

Since you should never be working alone, help should be within shouting distance. And in many cases, as with small trees, the only help you need to free the saw and fell the tree into the lay is another pair of arms pushing against the trunk.

Option 2: Use a Wedge

Wedges are very effective in lifting the tree off your saw bar and into the lay. Your greatest challenge, however, will be finding an adequate opening in the back cut to even set a wedge. Aluminum or steel wedges are better suited for this purpose, as they can withstand the abuse of being driven into a closed cut much better than plastic wedges. Oftentimes a suitable opening can be found just behind the saw bar (the side opposite the notch). In this instance, two wedges can be used, one on each side of the trunk.

Another trick is to cut a gently sloping 2-3-inch deep notch on the topside of the back cut using another chain saw or an ax (see above). This notch will make it easier to insert the felling wedge into the back cut. Take care not to drive the wedge into the stuck saw bar.

Option 3: Use a Push Pole

A push pole allows you to push at a higher position on the tree than what you could reach when pushing with only your hands. Push poles can be made in the field by cutting a 12 to 16-foot long section from a small tree. Prop the pole against the trunk (a 45-degree angle is best) and then push.

Option 4: Cut a New Notch

Another effective method is to use another chain saw to cut a new notch on the same side the original one was made (fig. 1), or on the opposite side of the tree, the side it is now leaning toward (fig. 2). The latter method is only possible if the site permits felling the tree in this new direction. Remember, felling into a new lay will require planning a new escape route as well. With both methods the notch opening should be cut a distance of one tree diameter above the first notch you made. Use felling wedges this time and set them as soon as possible.

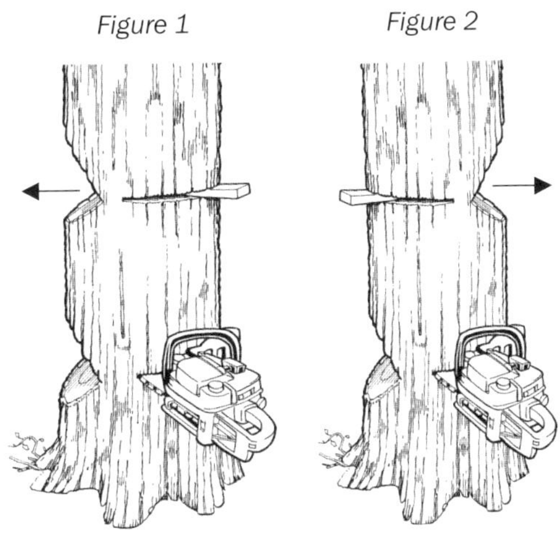

Recutting the notch: in the original lay (1), or in a new lay (2).

Option 5: Use a Pull Line

If the tree is large or has a heavy set back, a pull line may be the only safe option. Use the methods of rope installation described on page 101 and pull line operation described on pages 134. On set back trees that are more precariously balanced, take care that the exertions and forces of rope installation do not cause the tree to fall before you are ready.

Option 6: Get More Help!

When all options fail, it's time to get more help. Use a mobile phone if one is nearby. If you must leave to get help or send someone else, someone must stay behind to warn others of the hazard. If necessary, mark off the area with flagging and safety cones.

OPERATING A PULL LINE

You already learned how to install and anchor a pull line (pages 101-106) now you need to learn how to operate one to pull a tree into the lay. For felling trees with an unfavorable lean or into the wind, and avoid a set back tree, it's hard to beat the superior leveraging force a properly positioned pull line offers. If you haven't done so already, estimate the tree's height (page 86) so that those operating the line will know where they can stand safely. Standing slightly more than one tree length away from the tree should, in theory, be a safe distance to avoid being struck by the tree—provided your calculations are correct. To eliminate any doubt, stand at a distance of 1½ tree lengths, or more, away from the tree. Workers who are not directly involved with the felling operations should stand at least two tree lengths away from the tree.

Before the felling cuts are made, take up the slack in the rope and apply enough tension to hold the tree in position. Care must be taken that excessive force is not exerted on the line too early, stressing both the rope pullers and the tree being pulled. When cutting begins, it's important that you maintain verbal and visual communication with the cutter, regarding how much tension is needed on the line and when to apply it.

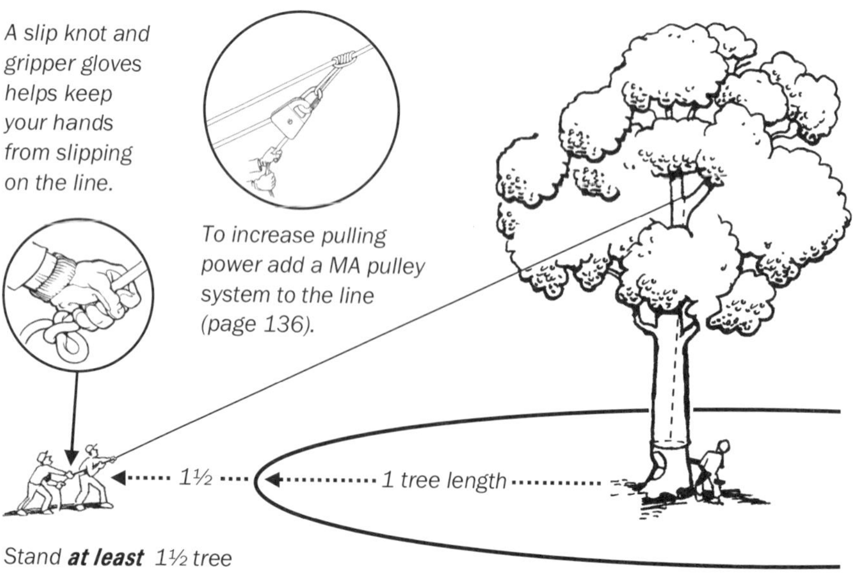

A slip knot and gripper gloves helps keep your hands from slipping on the line.

To increase pulling power add a MA pulley system to the line (page 136).

Stand ***at least*** *1½ tree lengths away from the tree being felled.*

MUST KNOW KNOT: **Butterfly Knot**

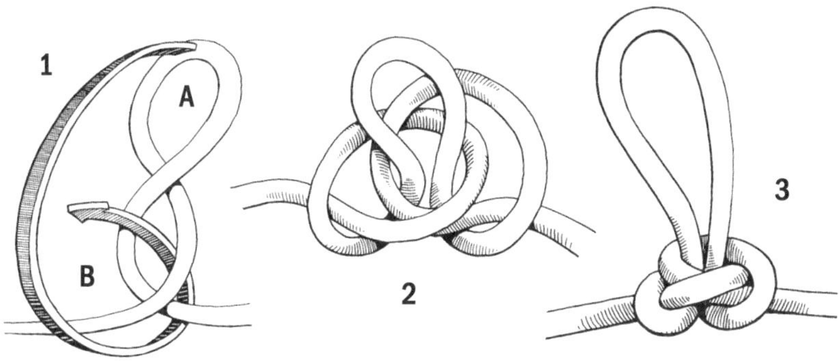

How to tie: *(1) Make a 360-degree twist in a bight of rope, forming an eye (B). Bring the end of the bight (A) downwards...(2) and pass between the legs of the standing part and up through the eye. (3) Pull on the end of the bight and both standing parts to tighten.*

Tying tip: *Take a large bight to form the 360-degree twist in the rope (fig. 1). It is difficult to pass the top of the bight (A) through the bottom eye (B) if the openings are too small.*

To keep your hands from slipping while pulling on the line, wear latex dipped "gripper gloves," or tie a slip knot just behind the rear pulling hand (inset previous page). When two or more workers are pulling on the line, use butterfly knots instead (see above), to prevent the knots from pulling out. The loop in a butterfly knot also functions as a convenient handle for pulling. Never wrap the rope around your hand or any other part of your body, which will keep you from being dragged away if the tree falls out of control! And for the same reason, keep the rope clear from your feet. You can also increase your pulling power by bracing your feet against a stump or pull from behind another tree.

The tree's size and amount of lean dictate the number of people needed to pull on the line or if a mechanical advantage system should be added (see next page). When the tree cutter announces the final warning that the tree is beginning to fall, or when it becomes plainly obvious, the pull line operator(s) should retreat along a planned and cleared escape route—one for each person pulling on the line.

MUST KNOW KNOT: **Prusik Hitch**

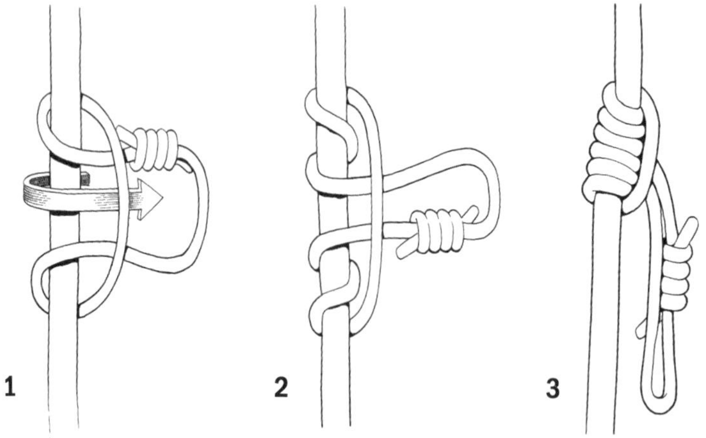

How to tie: *(1) Begin by making a girth hitch around the rope with a Prusik loop. Position the double fisherman's knot so it won't interfere with knot tying and carabiner attachment. (2) To form a 3-wrap (6-coil) Prusik hitch (considered minimum in most instances) take two more turns with the Prusik loop through the center of the original girth hitch. (3) Dress and set the hitch. Make sure the hitch grabs the line properly. If not, add more wraps with the Prusik loop until it does.*

Adding Mechanical Advantage (MA) to the Pull line

Large, tall, and heavy back-leaning trees often require a pulling force beyond that which one or two workers pulling directly on the rope can muster. In this case it may be necessary to add a mechanical advantage (MA) pulley system to the pull line. Simply put, MA is the factor by which ropes and pulleys multiplies the force, or effort, put into it. As it relates to our application here, MA is the measure of how hard you need to pull the rope to move a back-leaning tree. The inset on page 134 shows a three-to-one (3:1) MA system. Which means that the force exerted on the pull line is three times more than what would be offered without the MA system.

As a beginning ground worker, you probably won't be assembling a MA system on your own. But when asked, you should know how to tie a butterfly knot in the line (page 135) or attach a Prusik loop to the line with a Prusik hitch (see above), either of which functions as an attachment point for a pulley—the key component of a MA system.

Know Your Knots: Part 3

REPEAT, REPEAT, REPEAT. The key to tying knots exactly right every time is through repetition. And not just remembering how to tie them using your mind (*intellectual memory*), but remembering how to tie them in such a way that it is imprinted on your body as well (*muscle memory*). When tying knots becomes second nature, they can be tied even when you are distracted, forgetful, or fatigued.

Arborist Peter Jenkins says it like this, "when you tie a knot over and over and over again, your hands just sort of know what to do. Let me tell ya, when you get tired up in the tree [or on the ground] you don't want to depend on your mind so much, but the hands—you want them to know."

One good way of putting this advice into action is by using a method I learned from knot expert Brion Toss: keep a 5-6-foot hank of 1/8th to 1/4- inch diameter rope in your pocket, purse, man-bag, or glove compartment at all times—it takes up little room. Then you always have a rope available to practice knots at odd moments—waiting in line, traffic delays, TV commercials, lunch breaks, etc.—or to amuse family and friends (especially if it's a rope trick). In addition, the rope will frequently come to the rescue in everyday practical situations—trunk tie-down, impromptu dog leash, fish stringer. (If your practice rope is a chain saw pull cord you will always have a ready replacement for a worn one.) If you need to remember which knots to practice refer to the list of "must know knots" on page 92.

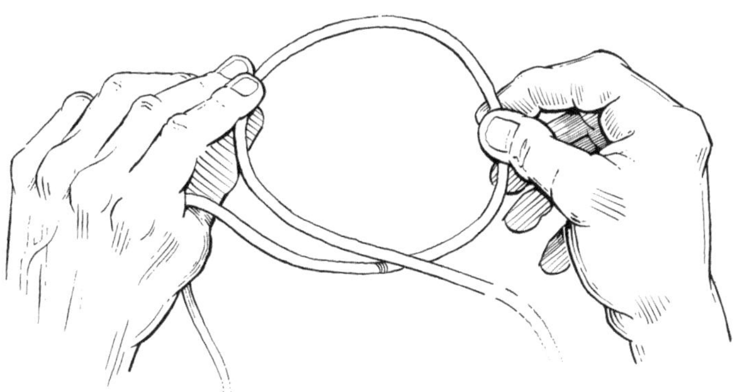

If you keep a short hank of rope handy you'll have the material to practice knots as well as for everyday practical applications.

CUTTING UP LIMBS, LOGS, & DOWNED TREES

Now that you have learned the fundamentals of roping down wood, chain saw safety, felling trees, and using a rope to pull over stubborn trees, it's time to deal with the wood on the ground. And whether that wood be limbs, branches, and logs from pruning and removal operations or downed trees from storm damage or felling operations, the cutting methods and principles that follow apply to them all.

Depending on the work order specifications, some (or all) of this material may be chipped and hauled away or some might be cut into firewood sections to be left for the homeowner. But regardless of what is to be done with the wood, you'll need to know how to cut it up in smaller sections so it can be more easily moved by you and your crew or the homeowner. This is the first step of cleanup and disposal. But before you begin cutting you need to be aware of the hidden tension forces in wood, which to the unfamiliar, can cause frustrating interruptions or serious injury.

Understanding Tension Forces in Wood

Tension forces in wood occur from the forces of gravity and the pressure of the ground (or other object) pushing against the weight of downed limbs and logs or an entire tree. Knowing how wood responds to these forces, and knowing where the pressure is located in the wood, will determine where safe saw cuts can be made.

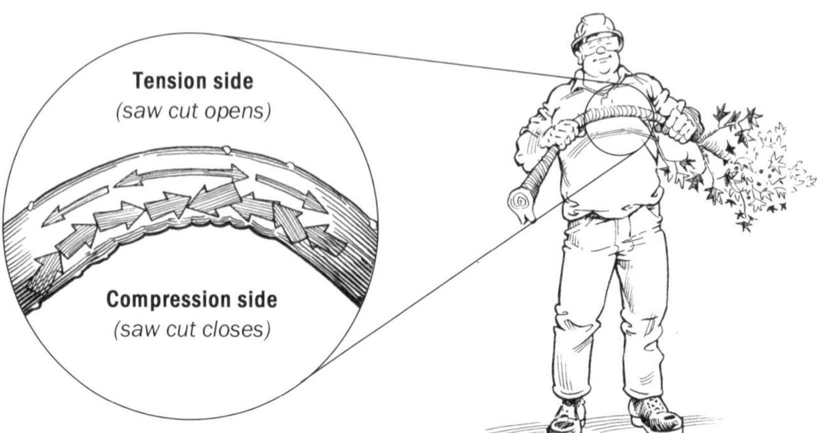

Tension forces found in wood

If you bend a small tree branch held between both hands (see illustration on page 138), the wood fibers on the outside of the curve (top) are being "stretched" or being pulled apart. This side of the branch is under "tension." Meanwhile, the wood fibers on the inside curve (bottom) of the branch are being pushed together. This side of the branch is under "compression." Of course, if you bend the branch in the opposite direction, these forces would be reversed.

Either way, limbing and bucking cuts made on the compression side always want to *close*, which often results in a pinched saw bar. Cuts made on the tension side, however, always want to *open.* Cuts made on this side first can potentially result in the limb springing dangerously toward the worker, as in the case of cutting *spring poles* (see page 140).

It will take experience to learn the variety of tension wood scenarios. A good rule of thumb though, is to remove small limbs under pressure with a single cut on the tension side. Whereas larger limbs under pressure should be removed in two cuts, as illustrated below. First, by making a shallow cut or small notch on the compression side (removing the saw before it closes the cut and pinches the saw bar) followed by a cut on the tension side. Since it is not always evident where the pressure in wood is located, it may be necessary to make light tentative cuts to see how the wood responds before committing to making the final cut.

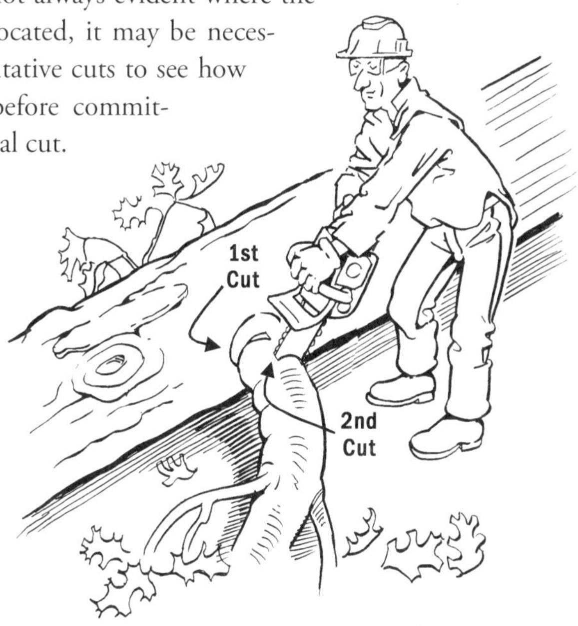

Remove larger limbs under pressure in two cuts: first, a small undercut on the compression side, followed by a finish cut on the tension side.

Spring Poles

A spring pole is tension wood on steroids. When cut, they can literally spring up and attack, seemingly out of nowhere, with a force capable of severely injuring or even killing an unwary ground worker. Spring poles are created when a falling tree bends a sapling over and pins it to the ground under tension. Rather than trip these booby traps unexpectedly, learn to identify them, and either avoid them entirely or deal with them by following the procedure below.

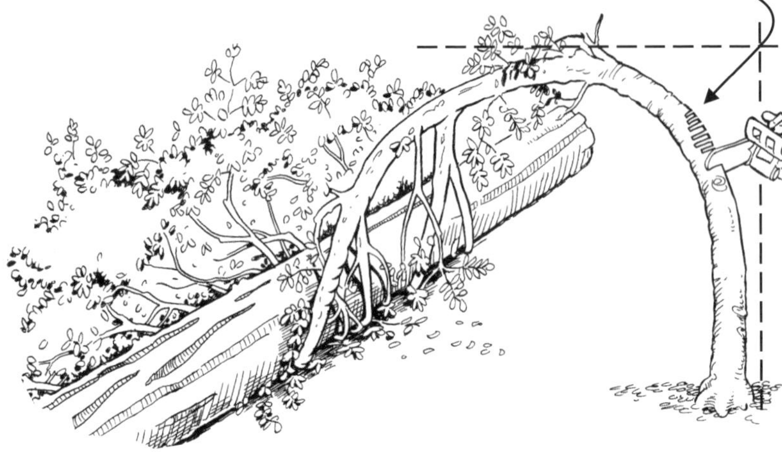

Releasing a Spring Pole (Below Shoulder Height)

1. Determine the maximum tension point of the spring pole. This is where you will begin making your saw cuts. Do this by first drawing an imaginary vertical line that extends upward from the base of the tree using your finger. Next, extend a horizontal line from the highest point of the spring pole until it intersects the vertical line. Finally, extend a line 45 degrees from the point where the two lines intersect until it reaches the spring pole. This is the point of maximum tension.
2. While standing at a 45-degree angle to the rear of the spring pole (to minimize your chance of being struck should it release unexpectedly), start your first cut at the maximum tension point on the top (tension side) of the spring pole.
3. Down from your first cut, make subsequent shallow cuts approximately one inch apart. The cuts should be made while running the saw at high speed. This will allow the fibers to break evenly as the tension is released gradually.
4. Cut the spring pole off once you see the tension release.

Basic Limbing Techniques

Technically, you don't limb a downed tree, but rather you de-limb one. Nevertheless, *limbing* is the term used to describe the process of cutting branches off the tree to make it easier to move or cut into shorter lengths. Though hidden hazards abound in this work, it is actually a simple task if performed thoughtfully and methodically. Each branch you approach is an adventure in problem solving as you determine if it is under pressure and decide where to make the cut. What follows are some basic rules and methods for removing limbs and branches.

1. ***Remove the pull line***—If a pull line was used to fell the tree (see page 134), untie and pull it free from the tree to prevent it from being cut.
2. ***Locate and release spring poles***—It is better to locate these ground hazards and release them before cutting rather than mistake them later as a limb.
3. ***Block the tree***—Block the tree if you suspect it could roll when supporting limbs are cut free. Remember, work from the uphill side of the tree whenever applicable.
4. ***Keep your distance***—It is always a dangerous situation when two or more people are working on the same tree. First, a worker could be struck by another worker's saw if they are working too close to each other. Second, the tree could shift or roll unexpectedly from a cut made by one of the other workers. Therefore, always know where the other workers are, be alert to what they are doing, and stay at least 10 feet away from each other.
5. ***Where to stand***—Try as much as possible to keep the trunk between you and the chain saw while cutting. The trunk functions as a barrier against cutting your legs or feet with the chain saw. If you need to position yourself on the other side of the trunk, apply the chain brake first.

As much as possible, keep the trunk between you and the chain saw.

Where to stand (continued)—In many instances it is more efficient to remove the limbs on the same side of the trunk on which you are standing, rather than to move to the other side. When doing so, pay careful attention to the position of the saw bar to avoid cutting your legs or feet! If possible, cut away from your body which will lessen the chance of this happening. Finally: never walk on or straddle the log to make limbing cuts.

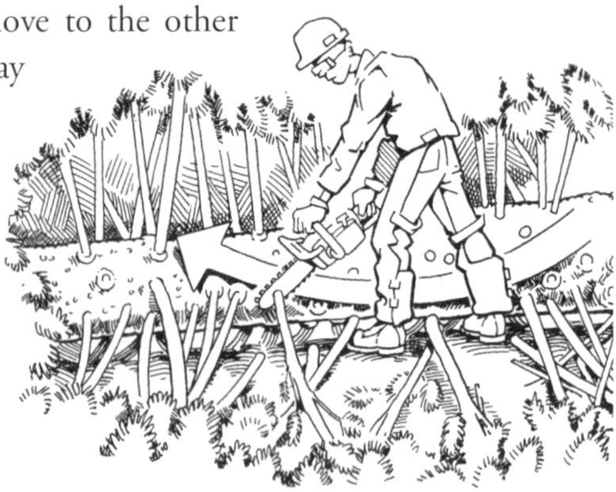

Cut away from your body whenever possible.

6. ***Cutting the limbs***—As you approach a limb, try to determine if it is under pressure or not. Small- and medium-sized limbs not under pressure can typically be removed with a single cut from either side. Limbs that are under pressure can be removed by using the one-cut method for smaller limbs and the two-cut method for larger ones (as described on page 139). Cut limbs and branches as close to the trunk as possible. This is done so it will roll, drag, and stack easier as you move and buck the wood later on. It sometimes is easier and safer to cut a larger limb under pressure farther away from the trunk, releasing the tension in wood that is smaller in diameter, and cutting the stub off flush afterward.

   Larger limbs (not under pressure) suspended in the air on the top or side of the trunk can be cut the same as you would a small tree—by first making a shallow, open-face notch on the compression side, followed by a back cut on the tension side. Cut the remaining stub flush with the trunk.

Fell upright branches like small trees.

If any branches are to be saved for firewood, cut them to length while they are suspended off the ground. This allows you to work at a more comfortable height while reducing the chance of hitting the ground with your chain saw.

Cut branches to firewood length while they are suspended off the ground.

Removing branches from the underside of the trunk is a bit trickier than cutting top and side branches, as the underside branches are the ones under the greatest pressure. The safest way is to cut the bottom branches last, after you have removed all the top and side branches. Do this by positioning yourself on the side of the trunk opposite the direction you anticipate it will fall, or roll, once you've removed the top of the tree. At this point you may need to push or use a cant hook to roll the tree over if it doesn't do so on its own. Keep the trunk between you and the side that has the majority of the branches as you remove them.

7. ***Moving the limbs***—If cut limbs accumulate to the point of interfering with your cutting progress or pose a trip hazard, move them off to the side. This will provide a better view of any other hazards or obstructions on the ground you wish to avoid. Remember to engage the chain break whenever you take one of your hands off the running chain saw or take more than two steps to clear limbs or move to the other side of the trunk. Shut the saw off if both hands are needed to move the material.

A final reminder—Things of great importance are worth repeating. In regards to limbing trees, it is this: *Pay attention to the saw bar!*—especially the tip—where it's at, what it's doing. You need to be watchful of those three dozen, or more, swirling knives as they come within inches of your legs and feet, encounter explosive pressure points hidden in the wood, and threaten to kick back in your face. Again, no one ever said tree work is safe work, but by simply paying attention you can help make it so.

Basic Bucking Techniques

Bucking is the term used to describe the process of cutting the main trunk and large branch sections into lengths for firewood, saw logs, or disposal. The two biggest challenges of bucking wood is doing it without getting your saw pinched or running it into the ground and dulling the chain. To avoid these problems study the examples in the pages that follow to make your saw cuts properly. Remember, saw cuts made on the compression side of the wood will close and cuts made on the tension side will open.

Bucking Wood Supported on Both Ends

A tree or log that is supported on both ends creates a compression force on the topside and tension force on the bottom. Here are two ways to buck a tree in this situation without pinching the saw bar.

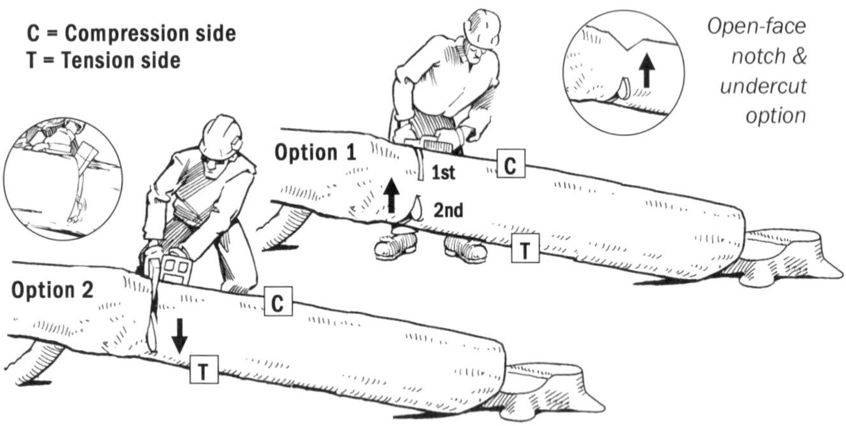

Option 1: The Two Cut Combo

Make a top cut (compression side) about 1/3 the way through the log. Finish with an undercut (tension side) so the two cuts meet. To release the log more slowly and with more control, make a shallow open-face notch on the topside and leave a hinge as you make the undercut (see inset above).

Option 2: Use a Felling Wedge

Begin by cutting from the top of the log or trunk and insert a plastic wedge snugly into the cut as soon as room permits and before the saw gets stuck. As you continue cutting downward, the wedge will prevent the saw bar from binding in the wood. It may be necessary to periodically drive the wedge in further to open the saw cut more.

Getting Attention

IF IT'S NECESSARY TO INTERRUPT workers who are running a saw, always approach them from the front to avoid surprise, or use a branch to gently tap them on the back or shoulder. Never try getting a saw operator's attention by throwing things at them, coming directly up to them yelling, or tapping them on the shoulder.

Bucking Wood Lying Flat on the Ground

One of the best ways a newbie groundie can get familiar running a chain saw, and with the tension forces found in wood, is by bucking up trunk sections lying flat on the ground. Doing so will also introduce you to another important tool: the cant hook. As you cut, your goal is to prevent the saw bar from binding in the wood and from making contact with the ground after it cuts through the log. Here are three methods to consider:

Option 1: Roll the Log

Cut 1/2 to 3/4 of the way through the log (as in fig. A) or at the first sign the saw kerf is beginning to close. Next, roll the log over with a cant hook, or by hand if it's small enough, to complete the cuts from the other side (fig. B). If you are cutting firewood, make your cuts along the entire length of the log section before rolling it. Be aware that dirt and gravel are often embedded in the bark of the bottom side of the log. Use an ax, stiff brush, or even a screw driver to clean the area you'll be cutting. You can finish the bucking cuts two different ways: you can cut the log from the topside so that the two cuts meet, or, as is my preference, insert the bar in the previously made cut from below and undercut the log (as in fig. B).

Option 1

Figure A

Figure B

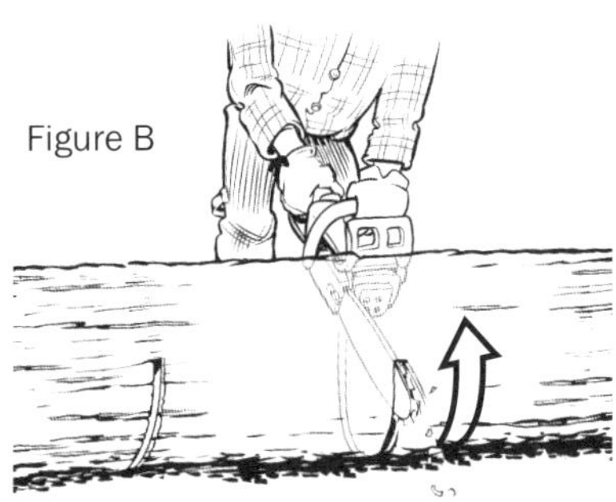

Option 2: Use a Wedge

Cut approximately halfway through the log from the top and insert a plastic wedge snugly into the saw cut. Continue cutting, but stop periodically to drive the wedge further into the log. As you do this, you should be able to see the log literally lift off the ground, providing the room you need to finish the cut without grounding the saw. Cut slowly as the saw bar nears and exits the bottom of the log.

Option 3: Use a Shim

Drive a piece of wood under the log where you intend to make the cut. A wedged-shaped piece cut from the end of a log works great. The wood shim will prevent the chain from making contact with the ground, as well as lessen any compression forces on the topside of the log.

Option 2

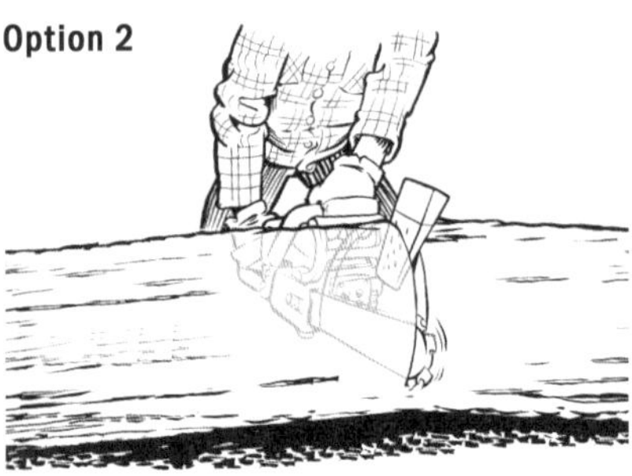

Option 3

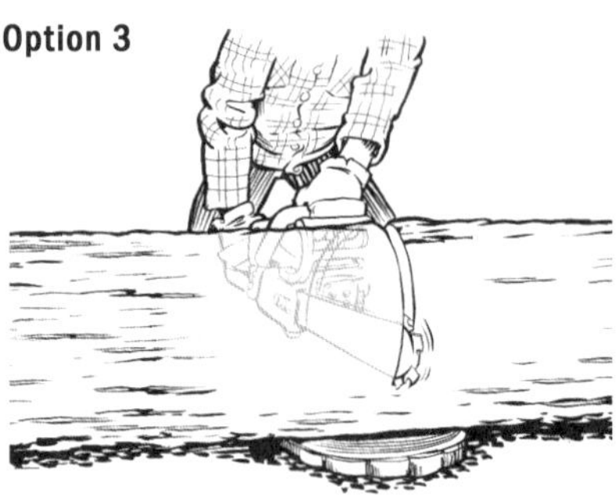

Bucking Wood to Firewood Length

The best answer to, "What is a good length to cut firewood?" is, "To a length that fits the stove or fireplace of the person burning the wood." More specifically, however, the wood should be cut to the length specified on the work order or by the customer, if the wood is for them (16-18-inch is considered standard). If in doubt, ask before cutting. Just as important as knowing the length, is being consistent in cutting the wood to that length. It is no fun stacking firewood of varying lengths. Use a tape measure, measuring stick, the chain saw bar, or chain saw body as a guide for consistency. Similarly, it is no fun trying to split wood that was cut at an angle. In fact it can be quite dangerous. Make your cuts perpendicular (90°) to the length of the trunk or limb.

Freeing a Stuck Chain Saw

It happens to everyone eventually. The quickest and easiest way to free a stuck chain saw is by grabbing another one, this time making the bucking cuts in the right place and taking the proper precautions. Sometimes, however, this results in two stuck saws. Then what? Well, before you yank the handle off your saw in frustration, try one of the methods below.

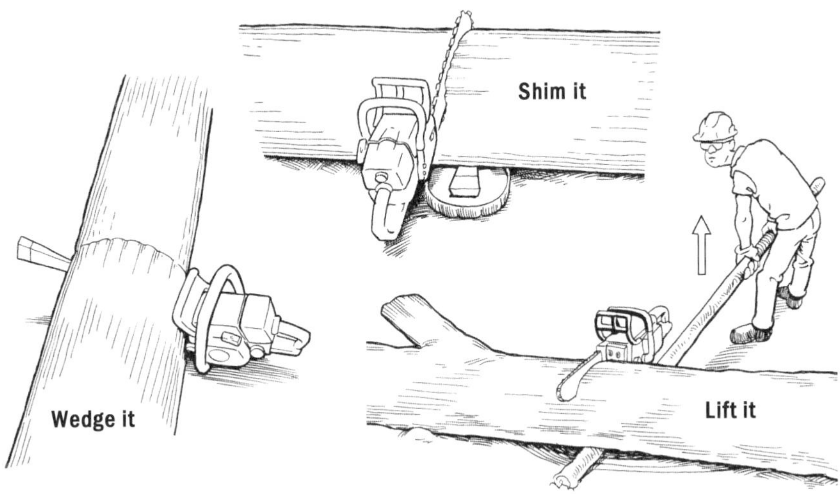

The Wedging Method—If the saw is pinched in the topside of the log, drive a wedge in the top, or more likely the backside of the log, with an ax or sturdy branch. You will probably have to probe around a bit for a big enough opening in the cut that will permit wedge entry. Be careful you don't drive the wedge into the bar and saw chain.

The Shimming Method—Force a piece of wood (small log, wood slice, large limb) under the log for the purpose of raising it, relieving the pressure in the compression wood, and opening the saw cut. Get even more lift by driving a felling wedge between the shim and the main log.

The Pry Pole Method—Select a sturdy limb or sapling to use as a pry bar to relieve the pressure exerted on the saw. If there is enough room, place the pole completely under the log and lift up on the pole. Or, if the pole can only be partially inserted, use a fulcrum, such as a small log or rock placed under the pole, and push down. Where you position the pole to get maximum leverage may require some experimentation.

MOVING LIMBS & LOGS

TDO—*take down only*—is a phrase some tree workers might be familiar with (it's certainly a favorite of our crew). It means: just get the tree on the ground—no disposal, no cleanup. The same for select pruning. Just leave the wood on the ground for the homeowner, or an unfortunate friend or family member, to do the cleanup. TDO jobs are the perfect end to a tough week of tree work. Unfortunately, TDO jobs are infrequent—nonexistent for some. The fact is, on most jobs limbs and logs need to be moved to the chipper or trailer for disposal and the property raked and cleaned up afterward. Most customers don't want to deal with the mess nor have the means to do so. For most ground workers this is the most strenuous and tiring part of their job. Granted, trucks, tractors, skid-steer loaders, mini-loaders, even horses for some, will make moving limbs and logs much easier than if they were to do it manually.

However, this equipment is not available to all tree companies, nor is it always practical to use. Terrain features, such as hills and wet and muddy areas, and site obstacles, such as fences, narrow gates, manicured lawns, and structures, make it next to impossible to get this type of equipment to the wood. In this case, good old-fashioned manual labor, coupled with some simple tools and techniques, is the only answer. What follows are seven ways you can perform this work safely and efficiently, while preventing an unwanted trip to the chiropractor afterward.

1. Lifting

Back injury occurs most often while performing the work of lifting, carrying, and dragging wood or any heavy object, from one location to another. As a result, it is the leading cause of lost work time for the tree care professional. Sometimes injury can be prevented by simply getting help from a coworker or by cutting wood into shorter, more manageable lengths. It also helps if you are in good physical condition. But when the object only requires the effort of one person to lift, here's how to do it properly.

1. ***Plan ahead*** before lifting. Know where you are going with the object and that you have a clear path there.
2. ***Examine the object*** for sharp edges, slippery spots, nails, or other potential hazards you want to avoid grabbing.

Basic Lifting Method

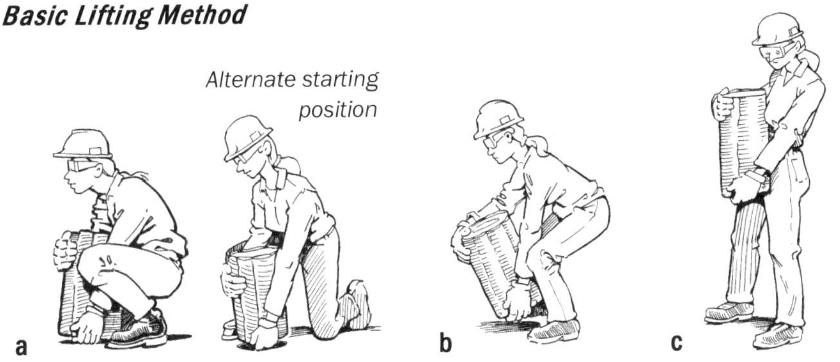

3. ***Squat down close to the object*** with feet about shoulder width apart (a). An alternate position is to start with one knee on the ground. Test the load to see if it can be safely lifted.

4. ***Grip the object*** firmly and begin slowly lifting with your legs (b), not your back, holding it as close to your body as possible. Keep your head up as you lift while maintaining normal back posture (c). Never twist your body during the lift, but rather turn by taking small steps after you have stood up straight. If you are straining to lift the object, it is too heavy. Set it down and get some help or simply cut the wood smaller. You might also try a different method entirely, such as using a log dolly (page 154).

End Over End Method

Wood sections that are too long or heavy to lift can be effectively moved by toppling them end over end. Begin by squatting down at one end of the log. Then lift the end by straightening your legs, keeping your head up and your back naturally curved. Finally, raise the log with your arms until it is nearly vertical before giving it a good push forward. Repeat the process until the log reaches its destination.

2. Carrying

Sometimes the most efficient way of moving wood and brush to the chipper or trailer is to carry it by hand. This is a standard practice when working in residential areas where you wish to avoid damage to property with groomed lawns, landscaped areas, and narrow gates. Also, carrying wood and brush keeps it from collecting dirt and mud that can dull the teeth of the chain saw and blades of the wood chipper.

The Shoulder Carry

1. ***Stand the log on end*** using the end over end method on page 149.
2. ***Put your shoulder on*** the balance point of the log. Tentatively lift the log to determine if it is positioned too high or too low.
3. ***Straighten up,*** using the legs, and using your arms to help lift and steady the log onto your shoulder. The log will almost lift itself horizontally into position if it is slightly back-heavy. Toss the log off to the side for easy unloading.

3. Dragging

As a ground worker, you will quickly become familiar with this activity as you will spend a tremendous amount of time doing it. There's not much to it really, but knowing a few tricks will make it easier and thus more tolerable. First, don't waste time and irritate your coworkers by dragging a single limb to the chipper, as many beginning groundies are prone to do—unless, of course, it is a large one. Instead, drag an entire pile of limbs by making them into easy-to-drag "sleds" (below left). The bottom framework of the sled is formed by placing the longest and widest branches together on the ground first, with the butt ends facing the direction the pile is to be dragged. Additional branches are tossed on the pile in the same manner. Place the smallest branches and debris on top of the pile last. At this point the whole pile can be dragged away by grabbing one or two of the "handles" that the bottom limbs provide.

Two clever ways to drag brush:

Secondly, using simple tools such as a webbing sling (above right) or hand tongs (below right) can also make dragging limbs and wood easier. A webbing sling girth hitched around the butt ends of a small brush pile makes a fine handle for one-handed dragging. Hand tongs allow you to grab, lift, and drag larger limbs and smaller diameter log sections with only one hand. Another benefit of using hand tongs is that you don't have to bend over so far in order to reach the wood. To some degree, all methods of dragging can cause dirt or mud to be ground into parts of the log or brush being dragged across the ground. To prevent dulling saw chain and chipper knives, clean off the affected area before you begin cutting or chipping.

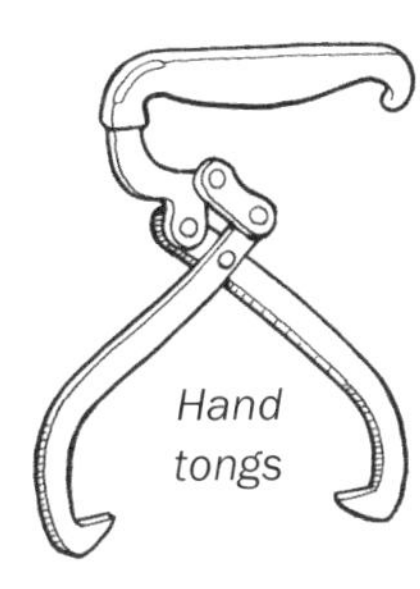

Before you begin dragging limbs and brush to the chipper, clear the drag path of obstructions and move anything that could get damaged: sprinklers and water hoses, gutter downspout extensions, lawn ornaments, lawn furniture, bird feeders, your own equipment. It is also important that you exercise extreme caution when dragging limbs—especially long ones—and brush piles through narrow gate openings or close to buildings. These, along with other objects, can be easily damaged when brush is dragged across them. In these instances it may be necessary to trim longer branches or drag (or even carry as shown on page 161), smaller brush piles.

Don't Waste Your Trips

BACK IN THE DAY when I was a waiter in the big city, I learned the importance of not wasting my trips. From the kitchen to the dining room and the dining room to the kitchen, I always had something in my hands. To the dining room I carried meals, a pitcher of water for refills, menus, the bill—always something. Back to the kitchen I carried dirty or unused plates, a new meal order, a dirty napkin; I never returned empty-handed. I quickly learned that not wasting trips moving about the restaurant was the key to efficiency and good service—and good tips.

The same holds true for the ground worker. With so much work to get done and limited time to do it, each trip from the work truck and chipper to the work area and back again adds up over the course of the job. So how can you make the best use of your travels? By anticipating what comes next. As you're dragging a pile of brush to the chipper, be thinking, "What can I bring back to the work site that we need, or will need eventually?" Perhaps it's a bigger chain saw and pull line to fell the spar, or maybe it's the rakes, dolly, wheelbarrow, or tarps needed for the eventual cleanup process, or something as simple and thoughtful as a water bottle for someone who needs a drink. In the same way, don't waste your trips to the truck. Take a quick look for something to bring back with you: a tool no longer needed, a coworker's discarded coat, a random lunch box, even a stray twig if nothing else.

To some, this sort of efficiency might seem fanatical. Instead, it actually becomes a sort of game—a game of awareness and economy of movement. You might not get a tip for all your efficiency, but it might help you and your crew get home at a reasonable hour.

4. Rolling

Moving wood by rolling it seems like a no-brainer. Yet it is amazing how frequently workers overlook this simple option. Instead, they make exhausting attempts at moving and lifting heavy log sections by hand, risking back injury in the process. By simply tipping the log onto its side, it becomes a wheel that can be rolled with your hands, feet, or cant hook to the desired location.

The more perfectly round a wheel is, the easier it is to roll. The same holds true with a log. You can't change the shape of the log (not much anyway), but you can remove any obstructions on it that could hinder it from rolling efficiently. This is another good reason to cut limbs off as close to the trunk as possible. Whenever practical, take advantage of slopes and roll log sections downhill. Do this in a controlled fashion, however; a runaway log is difficult to stop, and can cause serious damage to property or injury to people.

> "I once saw a man load a firewood log big around as a barrel by cutting it into slices right where they lay. His truck was downhill and he rolled the slices down there like huge hockey pucks, then up a plank into the truck body. Each slice was fully as much as he could handle, but he loaded them."
>
> – D. Cook
> *(The Ax Book)*

The cant hook is an indispensible tool for rolling and moving logs—your body will thank you every time you use it. The hook grabs hold of the wood while the long handle provides a mechanical leverage advantage, which makes log rolling significantly easier and more controlled. As described on page 145, the cant hook is used for rolling logs when bucking them into lengths. Longer, banana-shaped logs are extremely difficult and dangerous to roll by one person, as are logs being rolled uphill, or onto a trailer. Having two people using separate cant hooks is safer and easier in these instances.

The indispensible cant hook

5. Log Dolly

By definition, a wheel is a type of simple machine once it becomes fitted with an axle. As a result, the transportation of an object not only becomes possible, but is accomplished with amazing ease. That is the beauty of the log dolly, as well as the wheelbarrow. Once you've discovered the power and versatility of the log dolly for moving wood (and gear) you'll be spoiled as you try to make everything fit on this machine.

Though several models of log dollies are on the market, the ones best suited for this type of work are built sturdily, and able to handle loads up to 1,500 pounds, which is more than one person can move. They should have a curved back with wide, pneumatic tires. These offer good flotation on soft ground and are narrow enough to get through a standard 36-inch gate opening. In some instances, it may be necessary to secure the log to the dolly with rope or a ratchet strap to keep it from falling off.

The dolly in action on big wood.

6. Wheelbarrow

The importance of a wheelbarrow as a means for moving wood is best appreciated when experienced rather than described. Nevertheless, for anyone who needs convincing that wheelbarrows are for more than hauling dirt and concrete, consider the applications for the tree worker: it can be used to (1) haul gear to the felling site, (2) haul out the wood you cut—firewood, short logs, brush, and when a fine cleanup is necessary, the final rakings of tree debris, and (3) as a comfortable chair with backrest when it's time to take a break and admire your work.

The versatile wheelbarrow.

Wheelbarrows for tree work should have at least a 6-cubic foot capacity and wide pneumatic tires. As with a log dolly, these types of tires enable it to move over rugged terrain more easily and over soft ground without leaving ruts. Models with two front tires can haul even heavier loads and provide greater stability.

A Wheel Improvement

HAULING WOOD AND GEAR with dollies and wheelbarrows can be made more efficient with a few simple tricks.[19] First, you can use sheets of plywood (2 to 4-foot wide by 8-foot long) to create a smooth path when wheeling over soft, uneven, or muddy ground. Second, when going up an incline with a wheelbarrow, a coworker can assist by pulling on a short rope attached to the front end. The rope can be taken off once you've reached level ground. Lastly, to ensure the best performance from your wheeled machines, maintain proper tire pressure. Keep a tire gauge and bicycle pump on the truck so you can service them when needed.

7. Winches

A winch is an extraordinary tool, not only for moving sizeable logs or piles of brush, but for pulling out hung-up trees or a stuck work vehicle. Compared with come-alongs and block-and-tackle systems, winches have a much greater pulling force and can move the loads quicker and with less operator effort. Though truck-mounted cable winches are the most powerful, the portable, gas-powered ones are the most versatile. They can be carried by hand to where they are needed and operated anywhere a suitable anchor is available (tree, rock, etc.). The most popular and practical portables are equipped with a capstan winch that a rope can be wrapped around. Unlike winches with a fixed length cable, these winches allow for unlimited rope length.

Portable winches can be operated anywhere a suitable anchor is available.

More Ways to Move Wood

The methods of moving wood could fill a separate book because there are so many different work situations. Tree companies working in snowy conditions often use thick polyethylene plastic sleds to haul wood from the work area. Crews working in lake country will, just like beavers, fell trees directly into the water. From there they will either winch them up to the shore to be disposed of, or tow them away by a boat to a landing where the wood can be loaded or chipped there. No matter what the challenge, a solution always exists. Figuring it out is half the fun. Coming up with the best strategies requires imagination, creativity, and resourcefulness.

Staging Wood

In most instances, the limbs and logs that you've been lifting, carrying, dragging, and rolling end up in the chipper or on a trailer. And because there usually is so much of it (especially with removals) and it takes up so much room, thought must be given as to how to arrange, or "stage," the material until it comes time to deal with it. Limbs and brush in particular need to be piled in such a way that it is easy for the wood chipper operator to grab and feed it into the machine. The standard for this is to face the butt ends of limbs toward the chipper infeed chute.

Stage brush with the butt ends facing the chipper infeed.

Don't stage limbs so close that they block access to any parts of the chipper or become a trip hazard while working. Also, it is much easier to chip several small piles of brush than having to pull apart one big "beaver lodge" size pile before chipping. As the piles grow and the work area becomes more crowded, take the time to chip what you have before you deliver more. Before you chip, be sure to ask permission or alert the climber and others first.

CHIPPING WOOD

As you probably already know, the wood chipper, along with the chain saw, is one of the most dangerous pieces of equipment the ground worker will operate. It injures and kills scores of workers every year. If you've ever been to an arborist conference where Dr. John Ball, a Professor of Forestry, warns his viewers of an upcoming slide, it's likely he showed a picture of a wood chipper fatality. What you see is so bad that grown men cover their eyes. You cannot have too much respect and healthy fear for a machine that literally eats people alive.

But the wood chipper is also the tree worker's best friend. There is nothing like it for turning mountains of limbs and logs into a compact pile of useful chips. Besides that, it's fun to operate. But only when it's done safely by following the correct procedures and taking the proper precautions, along with receiving hands-on training from a qualified arborist. Refer back to pages 51-53 for general inspection guidelines and hook up procedures. For a fuller treatment of both, along with chipper maintenance, consult the operation manual specific to the chipper you are operating.

You cannot have too much respect and healthy fear for a machine that literally eats people alive.

Positioning the Chipper

- If the chipper will be parked on a public roadway, place the appropriate number of cones spaced at the proper distance apart to alert oncoming traffic and pedestrians (see page 68). If necessary, post a flag person to keep people away from the work area. The flag person, along with the chipper operator, should wear reflective or high visibility clothing.
- Never position the chipper under the tree you will be working on. You can always back the work truck and chipper to the brush pile afterward if necessary. Refer to page 62 for back up instructions.
- Leave enough room behind and on the sides of the chipper so the operator has plenty of room to chip brush and service the machine if necessary.
- When chipping with the machine attached to the towing vehicle, turn the vehicle engine off, engage the emergency brake, and chock the tires. If the chipper is to be detached, always chock the wheels first.

Dressing for the Occasion

In addition to wearing the personal protection equipment (PPE) described on pages 25-30, here are some dress reminders that are particularly relevant to operating and feeding a wood chipper.

- Wearing a helmet and safety glasses when running a wood chipper is a no-brainer (and required), but so is wearing a face shield attached to your helmet, even though it isn't mandatory.
- Operators should always wear work gloves, but not the gauntlet type with wide cuffs, which could catch on brush and pull you toward the chipper infeed. For the same reason, don't wear loose fitting clothing, saw chaps, jewelry, or have long hair uncontained while chipping.
- Chippers are loud. Some are REALLY loud, so much so that many workers wear foam ear plugs under their ear muffs to add extra hearing protection—not a bad idea when operating any noisy equipment.
- Wear sturdy work boots. You'll need the sure footing they offer when dragging piles of brush over rugged terrain on your way to the chipper.

Starting the Chipper

1. ***Before starting*** the chipper make sure that:
    - You get the climber's permission and inform your coworkers that things are about to get noisy.
    - The clutch is disengaged.
    - The discharge chute is pointed in a safe direction.
    - The chipper hood lock pin or bolts are secured.
    - The infeed hopper is free of tools or foreign objects (something we learned the hard way when a leaf blower got chewed up).
    - The feed bar is in neutral (on hydraulic feed chippers).
    - All workers are clear of the machine.
2. ***Start*** the chipper engine according to the operator's manual and then:
    - Allow the engine to idle and warm up for a minute or two.
    - Slowly engage the clutch.
    - Increase the engine RPM to operating level.

Chipper Anatomy

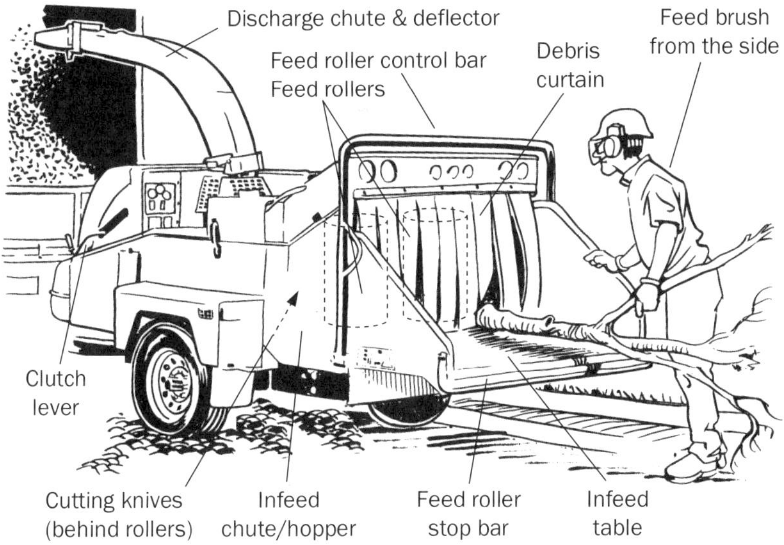

Feeding the Chipper

Back in chapter 1 you learned the three-step learning process—*tell me, show me, let me*. As important as the first two steps are, it's the *let me* step (also known as personal experience) that will teach you best about what to do and what not to do when chipping wood. Who could ever forget the horrifying experience of being pulled toward chipper rollers while trying to free a gloved hand caught on a limb? You can read about it and be warned about it, but when it happens you never forget. Experience can be a cruel teacher, but also a highly effective one. Nevertheless, the more you know beforehand, the better. Here are some beforehand tips:

- As often as possible, feed material into the chipper from the side of the infeed chute as shown in the illustration above. The reason is two-fold: one, to keep the operator from being struck by the material as it often twists and turns its way through the feed rollers, and two, to prevent the fed material from pulling *you* into the feed rollers. If you're working on a street, feed material from the curbside to avoid stepping into traffic or being pushed into traffic by the material being fed into the chipper.

- Follow the four step method of feeding material into the chipper: (1) feed the butt or cut end of the brush or log into the chipper first, (2) release the material as soon as the feed rollers grab hold, (3) turn away from the feed table, and (4) keep moving to get more material to chip.
- Use a long branch to push small material into the feed rollers, or simply feed a long branch into the rollers to push the material through.
- Keep a chain saw nearby to reduce large diameter wood and large limbs into more manageable pieces to feed into the chipper.
- As stated earlier (page 115): ***look out for ropes!*** so that they don't get accidentally gathered up with brush and thrown into the chipper.
- Examine brush piles for stones, wire, nails, or other objects that could damage the chipper knives and become dangerous projectiles.

Things Not to Do

If you dutifully obey the five *nevers* that follow, you'll likely avoid being an unfortunate victim like the one below or those depicted in Dr. Ball's slides.

1. ***Never*** reach beyond the back edge of the infeed chute/hopper with any part of your body to feed material into the feed rollers. Also, never assume the feed roller control bar or stop bar will, in fact, stop the feed roller or reverse its direction.

This warning decal says it all.

2. ***Never*** climb onto the feed table to unclog or clean the infeed chute/hopper while the chipper is running.
3. ***Never*** perform maintenance on the chipper unless the engine is turned off, the ignition key is removed, and the cutter wheel has come to a complete stop. Also, never perform maintenance on the chipper unless you are qualified and authorized to do so.
4. ***Never*** use a rake, shovel, or other such tools to push material into the feed rollers or clear the infeed area while the chipper is running.
5. ***Never*** rely solely on any book or owner's manual to teach you how to safely operate a wood chipper. Get instruction from a qualified arborist.

CLEANING UP THE JOB SITE

The finishing touch to any tree job is a good cleanup. For many customers a good cleanup is what they comment on and remember most (along with a courteous and skilled crew, of course). "We didn't even realize you guys had been here until we saw the pile of cut up wood—everything was cleaned up great!" This, along with others like it, are the type of testimonials we often get from our customers. So, when your supervisor obsesses over a small twig you missed while raking, it's because the boss knows that an exceptional cleanup is one of the reasons why they repeatedly call you back to their property and proclaim your greatness throughout their neighborhood. Besides that, a good cleanup is satisfying in its own right, for it gives you a sense of pride and accomplishment in doing things well.

Raking

For most ground workers, the first thing that comes to mind when they hear the call for cleanup is *raking*. The other thing that comes to mind is tedious labor—boring, monotonous, and repetitious work. But compared to the physical rigors and hazardous nature of the rest of the work you've done thus far, raking can actually be a peaceful, relaxing, even mentally therapeutic activity. However, even this seemingly simple task takes skill if you want to do it efficiently. Here are some tips to get you started:

1. First off, raking should be one of the last cleanup chores that you do, *after* everything is moved out of the way: branches, logs, equipment.
2. One of the goals of raking is to rake any given area only once. Don't make the thoughtless mistake of raking debris or moving brush across an area already raked clean (a great way to become the most unpopular guy on the crew). Apply the *Basic Principles of Cleanup* described on the next page.

Think twice, rake once.

Three Basic Principles of Cleanup

THERE ARE THREE BASIC PRINCIPLES for job site cleanup that maximize job efficiency and minimize crew frustration: (1) clean up *from large-to-small,* (2) *from high-to-low,* and (3) *from far-to-near.*

From large-to-small means that you move the biggest wood from the work area first and clean up the smallest and finest material last. For one thing, smaller wood and tree debris are often trapped under bigger wood. Moving big wood first liberates the small stuff so you can see it and get to it later. Also, when moving larger wood, especially trunk wood, a trail of loose bark is the last thing you want deposited over the area you just raked up. There are, of course, times when the reverse of this principle applies, like moving limbs and smaller wood so you can better access the larger wood.

From high-to-low means that you clean off roofs, decks, and other elevated places before raking underneath them.

From far-to-near means that you begin cleaning up the job site at the outermost perimeter of the work area and work progressively inward toward the center where the chipper and work truck are positioned.

3. Rather than making a bunch of little piles of rakings, make larger piles (but not too big) that you can pick up with your arms, or better yet, a silage fork, to be put on a tarp or in a wheelbarrow for disposal.
4. Avoid unnecessary bending to pick up small debris piles. In the same way a broom and dust pan are used together, use two rakes, or one rake and a scoop shovel, to pick up the piles. This works best with two people.
5. By inverting a rake so the tines are facing up, you can push small debris piles to their destination rather than repeatedly rake them forward.
6. Never throw rakings that contain rocks, gravel, nails, wire, or other debris into the chipper that could dull the knives and become dangerous projectiles. Toss them in the back of the chipper truck, or, when working in more rural areas, into the woods and spread out with a rake.
7. Avoid leaving rakes and other hand tools on the ground where they can be stepped on or driven over and broken. Prop them up against a tree, fence, or building to avoid breakage and make them easier to find.

Other Cleanup Tools

Forks and shovels. While the rake may be the preferred tool for rounding up tree debris into their respective piles, the silage/pitch fork and shovel (and sometimes the rake) are best suited for picking up this material. In his classic *Arborist Equipment*, Don Blair identifies at least five different types of pitch forks used for picking up everything from small brush piles and pine needles, to wood chips and stump grindings. You may not need quite that many, but then again, if they're available, you just might use them.

Like the pitch fork family, there are countless varieties of shovels as well. A grain scoop shovel is a good one for all-around cleanup use, again, as a dust pan to sweep or rake debris into and a pointed spade shovel is good for digging out rocks before stump grinding. You will likely find all sorts of other applications for the shovel as the need arises: scooping up pet poop in the work area, filling in ruts made in the lawn with topsoil, shoveling wood chips under the tires of a stuck work truck.

Tarps and wheelbarrows. All the debris you've been raking, forking, and shoveling need to be put into something to be hauled away. You know from page 154 the merits of the wheelbarrow for this purpose. But did you know how useful poly tarps are for doing the same? Fact is, tarps are the most inexpensive, compact, and repairable (duct tape) tools for hauling debris (dragging or carrying) to the chipper, trailer, or into the woods (see also page 85). A tarp measuring 6x8 or 8x10 is a good size; anything larger than that will make it next to impossible to drag after you've loaded it up.

Brooms and blowers. The final touch of cleaning up is to use a broom or leaf blower to clean off roofs, decks, sidewalks, and driveways (even lawns when using a blower). The broom is usually reserved for small cleanup jobs, but can handle bigger ones if necessary. The leaf blower, on the other hand, is the best tool for these purposes—noisy as it is. A few words of caution are in order when using this tool. One, avoid blowing debris towards people, vehicles, and buildings; this has the same effect as using a sand blaster. Two, modern leaf blowers are powerful, so take care when blowing around delicate objects—small lawn ornaments, flowers, and other landscape plantings. Lastly, using a leaf blower on a roof is extremely dangerous. Don't go on any roof until you've learned what safety precautions should be taken and have received permission from your crew leader.

The Lost Art of Coiling Ropes

PUTTING ROPES AWAY is a common task of the tree worker. Whether done immediately after using them or later in the day while packing up to leave, ropes need to be picked up and stored. Traditionally, ropes were masterfully coiled into a work of art and hung in the work truck or shop to air out. Now however, ropes have gone into hiding, oftentimes housed inside unmarked mystery bags. Coiling rope is now largely viewed as a troublesome and outdated method of storing rope reserved for those who can't afford rope bags.

To be fair, rope bags have countless benefits which coiled ropes do not (see pages 109 and 115), which is why they are so popular and why my crew and I love them. Nevertheless, coiling ropes is still a necessary skill that the modern arborist needs to learn, if for no other reason than to hang a wet rope to dry. If the rope is also dirty and needs cleaning, having it coiled and hung acts as a visible reminder to do so. What follows are three ways to care for your precious ropes and show off your handiwork at the same time.

Gasket Hitch Coil

This is a good one to learn first, as it will serve as a good primer for learning the other two that follow.

(1) Coil up all but the last 5-6 feet of rope.

(2) Make several horizontal turns around the coils.

(3) Pass a bight of the remaining rope through the space above the turns. Open the bight, sliding it over the coils...

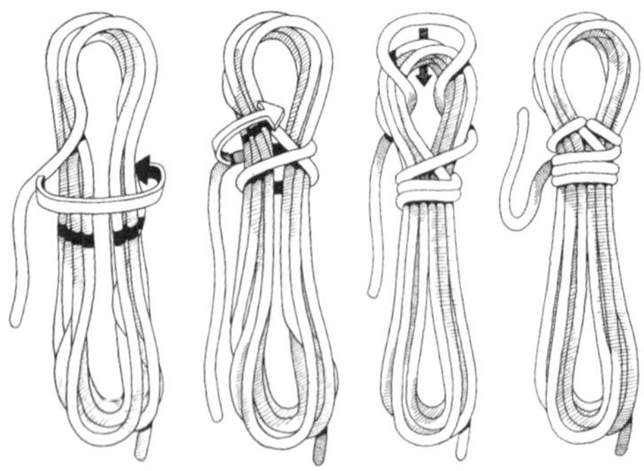

1 2 3 4

(4) ...down to the top of the turns. Pull the end to set. Hang the coil from the top loops or an appropriate knot tied in the tail.

Performance tips: When you want to employ the rope for a task, don't just throw the coil on the ground where it will quickly tangle, instead methodically flake it into a pile on the ground. In addition, a miniature version of this coil (12-14" long) can be used as a throwing knot for installing a rope in the tree.

Lineman's Coil

The lineman's coil is easy to tie and untie, and finishes with a convenient eye from which the coiled rope can hang.

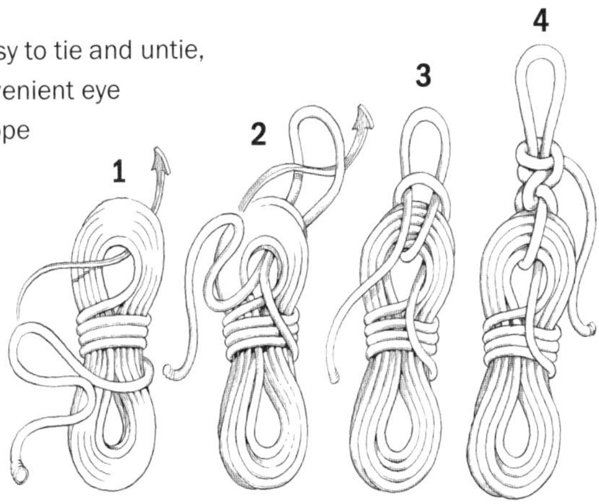

(1) Coil rope as usual (page 164) and form a bight near the end.

(2) Pass the bight through the space above the horizontal turns.

(3) Pass a second bight through the first bight. Then pull the slack out of the first bight and orient it to the back of the coil, so the hanging coil will lie flat against the wall.

(4) Adjust the second bight to the desired loop size and lock it all down with two half hitches.

Figure-8 Coil

The figure-8 coil is another method that has a handy loop to hang it by. Also, when coiling the rope doubled (as shown below), it takes half as long to coil and uncoil than when doing so with single strands.

Coiling tips: Locate the ends of the rope and coil it doubled, either "butterfly" fashion (below left) or as normal (previous page), leaving a 2 to 3-foot loop of rope for locking the coils off as in figures 1-3 below. Use the spread of your arms to gauge coil length and one of your hands to hold the coils. If it becomes difficult to hold the rope, coil it over your wrist instead.

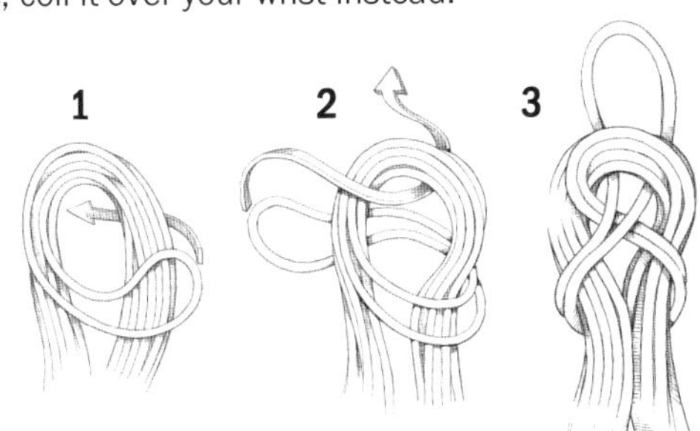

THE LAST THINGS

Congratulations! You've completed the job. And hopefully, you've completed it with the type of success described in the sidebar on page 69—no one hurt, nothing broken, everyone happy. But there are still a few more things to do before packing up and heading back to the shop.

Patch it Up

Once the work area has been cleaned up, it becomes much easier to spot such things as lawn damage from dropped wood pieces, ruts made from work vehicles or equipment, and broken pavers or cracked sidewalks. In the first instance, a tool, such as a four-tine spading fork, is often all that is necessary to "fluff up" the soil to its original form. In the latter three cases, however, more time, effort, and expense may be required to remedy those situations. It may even require a return trip with topsoil and grass seed, or new pavers that match the old. In either case, always inform your crew leader of any property damage, who in turn will inform the customer.

Put it back where you found it.

Put it Back

Make sure to put back any outdoor furniture, lawn ornaments, bird baths and feeders, clothes lines, etc., that you moved earlier. If you can't remember where it went, ask the homeowner. This simple courtesy is often overlooked by tree workers, resulting in needless inconvenience and even injury to the customer.

Many customers don't have the strength or manpower to put back heavier objects, but this doesn't stop some of them from attempting it anyway, sometimes wrenching their back or ankle in the process. And if the customer wasn't present while the work was performed and returns at night, they could easily trip over or run into an object unawares because it wasn't in its normal location.

Also, as you put back your own tools and equipment, make sure that they too are returned to their proper place on the work truck. Note also any repairs that might need to be made at the shop, or reported to your boss.

"Check Out" With the Customer

Just as you "checked in" with the customer when you first arrived, it's also important to "check out" with them before you leave. Inform them that the job is completed and you will soon be leaving. Ask them if everything looks satisfactory, if they have any questions about the job, or if they have (the dreaded) "just one more thing" for your crew to do. If they have any concerns or complaints about the work—especially in regards to property damage—that you can't answer, kindly instruct them to talk with either your crew supervisor on the job site or someone back at the main shop or office. Finally, thank the customer for the opportunity to work for them.

Checking out with the customer is typically the crew leader's responsibility, but everyone on the work crew should be able to do this properly and courteously. In the same way that you only have one chance to make a good first impression when you first arrive, you only have one opportunity to make a good lasting impression when you leave.

Do a Walk Around

After you've received a thumbs up from the customer and all the gear has been put back on the truck, it's time that someone (or two) do a "walk around." To do a walk around is to take one last look around the job site to make sure nothing is forgotten or overlooked. To come back for a forgotten tool is always a hassle and costly in time and dollars for someone.

You may have to search the work area carefully for some tools, especially ones left laying flat on the ground. But even big, bright, and obvious tools, such as ladders, wheelbarrows, and safety cones, are sometimes forgotten at the job site. And when debris such as lunch wrappers, drink bottles, or even a lone limb, is left behind, it's just one more thing the customer has to clean up and possibly even complain about (and for good reason). In the same way, do a walk around the work vehicle, chipper, and trailer to ensure that everything is hooked up correctly, logs and equipment are tied down properly, and tailgates and tool boxes are latched securely. At this point you are finally ready to leave the job site.

Leaving the Job Site

Unless you need to fuel up the work truck, dump chips somewhere, or head to another job, the next, and last stop, is the shop. To get there without incident will require applying the same principles of driver safety and passenger watchfulness that you learned back in chapter 3. Once again, our "One Thing" maxim—Pay Attention!—may be the deciding factor whether you make it back to the shop safely or not. Drivers: though you may be exhausted from a long day of work, don't let your guard down now—you still have precious crew and cargo to deliver to the shop—stay alert. Passengers: this is not the time to snooze or be preoccupied with your phone; help the driver stay awake—be conversant; help the driver see danger on the road—be watchful.

AT THE END OF THE DAY

After arriving back at the shop, it may be necessary to fuel up and sharpen chain saws, repair broken equipment, clean out the truck cab, or anything else to make things ready for the next day of work. It may also be necessary to inform your supervisor of any work injuries, customer concerns, or property or equipment damage that needs attending to. Whatever your end-of-day procedures are, make sure they get done before punching out.

At day's end most workers are eager to head out and go home, or do whatever it is they find pleasure in after a hard day's work. Perfectly understandable. Some, however, like to hang out for a while and talk about how the day went: the good (the flawless removal), the bad (the close call), and the ugly (the direct hit). Some companies even schedule time at the end of the work day, or work week, for these types of discussions.

> "I think we delight to praise what we enjoy because the praise not merely expresses but completes the enjoyment."
>
> –C.S. Lewis

Either way—on the clock or off—reflecting on the events of the day can be beneficial in a couple of ways. First, with the work day still fresh on everyone's minds, it's easy to recall the details of the day—what went well, and not so well. Valuable lessons can be learned from these discussions, which can be applied to future jobs.

Secondly, when coworkers exchange stories about work—or their life—it builds and deepens friendships, which in turn strengthens team unity. And when these conversations are seasoned with praise or admiration of a coworker's efforts, the day ends on a positive note, making the next day of work something to look forward to rather than dread.

FINAL REFLECTION

After running a tree service since 1989, my wife and I have been blessed to have had many wonderful people work for us. Some worked as a groundie for only a season and then moved on to something else. Others stuck around for several years, transitioning from ground worker to climber, before finding different work. Yet others have made tree work their lifelong calling—working for another company, starting up their own, or, as in the case of our son, carrying on the family business.

But, whether long or short, part time or full time, groundie, climber, or office manager, they all attained the same one thing: they became better versions of their former selves. They became better workers, better spouses, better parents, better friends—quite frankly, better people than when they first began. Work that is meaningful, relational, challenging, adventurous, achievement-oriented, and even dangerous, like the kind found in the tree care profession, has a way of doing that to people.

Tree work may be a "hard dollar" as many have said, but it is also a job, vocation, career or calling, that promises great reward for those who have what it takes. It comes as no surprise to me then, when I hear a former employee, or a current one say: *"Best job I ever had."*

APPENDIX A

A Woman in Tree Care

by Sharon Lilly

My career in the tree care industry began as did many others', working as a "groundie." At fifteen, I was underage, but had convinced the owner of the small tree service company near my home to let me work with them on Saturdays. I was drawn to it, mostly due to my love of climbing trees as a child, but also, inexplicably, to the hard, physical work. I guess I felt a part of a team as the crew pulled together to accomplish tasks that seemed so big to me. Working Saturdays soon expanded to working every school break and then every summer full time. Eventually, it led to a change in the trajectory of my life: a switch from pre-med to horticulture, and a future dedicated to tree care education.

From the start, I wanted to be a climber and I nagged continuously to be given a chance. I was told, as were most ground workers, that there were things I needed to learn on the ground first—how to tie the basic knots, how to work the ropes, what to prune and how prune it. Besides, climbing was deemed too dangerous for a girl.

Eventually, I was given the opportunity to climb and I learned as quickly as I could. It was almost magical for me. I loved figuring out how to access the most distant parts of the largest trees. I loved swinging among the branches. I loved the view of the world through the foliage from above. I even loved overcoming my fears when I had to do something that felt scary.

Because I worked for a small company, my coworkers had come to accept me as one of them. They saw how much harder I worked to pull my weight, knowing that I was not as strong as they were. But I did have to prove myself to every new employee who came along. And clients almost always assumed I was a boy—the owner's son—until they saw me up close. They were always surprised to meet a girl doing tree work, but the encounter often resulted in a tray of lemonade and cookies.

There were challenges along the way. At times, people made assumptions about my motivations or abilities, and there were clients who doubted my expertise. It seemed I was perpetually covered with scratches and bruises, my badges earned through hard work. Somehow, I never minded that much; it all came with the territory.

It seemed I was perpetually covered with scratches and bruises, my badges earned through hard work.

In my early years I never met another woman working as an arborist or tree worker. Even today, the percentage of women in the workforce who work on tree crews is exceedingly small. I believe, however, that there are many uncounted women who work in many capacities for tree care companies. I have met wives, mothers, sisters, and daughters who are part of family-owned tree services who contribute in myriad ways. Many do work in the field and some have become climbers. Often, they learn the science and best practices of arboriculture and eventually become ISA Certified Arborists®.

There have been many trailblazing women in arboriculture, and each has persevered despite the occasional headwinds and setbacks. I know researchers and educators who often had the "Dr." left off their names when their colleagues did not. I've met municipal and utility arborists who were the only women among their peers and who were typically asked to be the note takers at every committee or staff meeting. My friends who are climbers have persisted to get climbing gear, boots, PPE, and clothing that fit women better, just so they could do their jobs.

Certainly too, there are challenges for a woman in arboriculture to overcome. Long-held stereotypes and biases can present obstacles that must be surmounted. It takes courage and perseverance, at times, to keep going and not lose one's spirit. But the women I have known have all had one thing in common: their passion and determination to succeed.

For all people, men and women, a career in arboriculture is a journey that takes many twists and turns along the way. We learn from every setback and we grow stronger. We embrace the friendships that we forge over the years. And we celebrate the joys and rewards of working in, with, and for trees.

**For the first 25 years of Sharon's career, she worked in the commercial arboricultural field as a climber, tree service owner, and consultant. Eventually her love for education led her to the International Society of Arboriculture where she led the development of many educational programs, books, videos, and computer-based training for arborists.*

APPENDIX B

Two More Knots

After you've mastered the top ten "must know knots" you're ready to learn two more important knots: the ***cow hitch*** for anchoring a rigging sling to a tree trunk and the ***turquoise turtle knot*** (shoelace knot) for keeping your bootlaces securely tied.

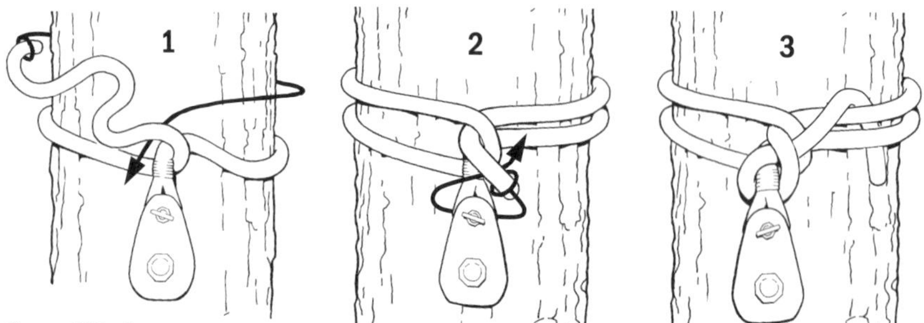

Cow Hitch

(1) Pass the end of the rope around the tree, under the standing part and back around the tree in the opposite direction. Pass the end through the bight formed by the two wraps. (2) Finish the knot by tying a half hitch around the "throat" of the sling. (3) Tuck the remaining tail under one, or both wraps of the hitch.

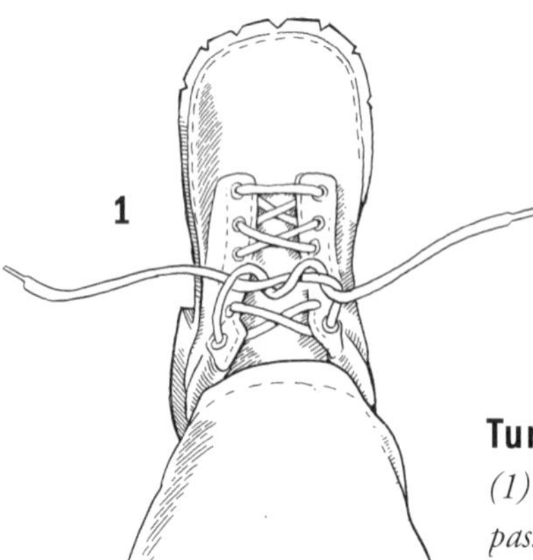

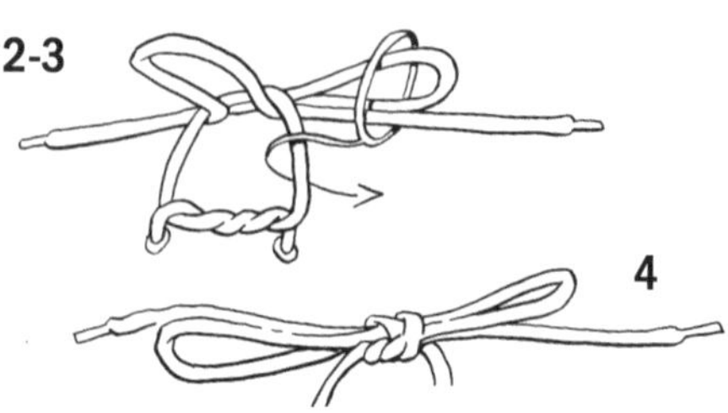

Turquoise Turtle Knot

(1) Start with your ordinary overhand knot, but pass one end around a second time to help keep the laces tight while you tie the rest of the knot. (2) Make the bow, or loops, as usual but leave extra space between them and the double overhand knot. (3) Pass the loop and end of one side through this space. (4) Tighten the knot as you usually would. Release the knot by pulling the ends.

APPENDIX C

Recommended Reading

If you want to go far in life, read books. If you want to go far in the tree care profession in particular, start by reading the books below. Although you can certainly learn a lot from videos, books (in print form) are generally more useful, durable, and, if you give them a chance, more engaging and enjoyable. Besides that, they smell good.

On Tree Work

Arborist Equipment (Second Edition). Donald F. Blair. International Society of Arboriculture, 1999.

Chain Saw Safety and Field Maintenance (English and Spanish Language Edition). Kevin K. Eckert. Hot Pixel Press/Arbor Global, 2013.

The Art and Science of Practical Rigging. Peter S. Donzelli and Sharon J. Lilly. International Society of Arboriculture, 2001.

The Fundamentals of General Tree Work. G.F. Beranek. Beranek Publications (www.atreestory.com), 1996.

The Tree Climber's Companion—2nd Edition (2000); *To Fell a Tree* (2009); *Knots at Work* (2013). Jeff Jepson. Beaver Tree Publishing.

Tree Climber's Guide (4th Edition). Sharon J. Lilly and Alex K. Julius. International Society of Arboriculture, 2020.

The Sibley Guide to Trees. David Allen Sibley. Alfred A. Knopf, 2009.

Tree Care Industry Magazine. A free monthly publication of the Tree Care Industry Association. View online at www.tcia.org.

On Teamwork, Leadership, & Character

Endurance: Shackleton's Incredible Voyage. Alfred Lansing. Basic Books, 2014.

Grit: The Power of Passion & Perseverance. Angela Duckworth. Scribner, 2016.

In the Kingdom of Ice. Hampton Sides. Anchor Books, 2014.

The Ideal Team Player. Patrick Lencioni. Jossey-Bass, 2016.

The Road to Character. David Brooks. Random House, 2015.

APPENDIX D

Glossary

Aerial rescue: method of bringing an injured worker down from a tree or aerial lift device.

ANSI Z133: in the United States, industry-developed, national consensus safety standards of practice for tree care.

Arboriculture (ar-bór-i-cul-ture): practice and study of the care of trees and other woody plants in the landscape.

Arborist: professional who possesses the technical competence gained through experience and related training, to provide for or supervise the management of trees and other woody plants in residential, commercial, and public landscapes.

Ascender: mechanical device that enables a climber to ascend a rope. Attached to the rope, it will grip in one direction (down) and slide in the other (up).

Barber chair: dangerous condition created when a tree or branch splits upward vertically from the back cut, slab up.

Basal anchor (BA): a method of anchoring a climbing or rigging line by attaching it to the base of a tree.

Big Shot: a large offset slingshot designed to fit into the end of an eight-foot pruner pole for setting throwlines in taller trees.

Biner: slang for "carabiner."

Block: (1) heavy-duty pulley used in rigging. Designed for dynamic loading (compare to *pulley*). (2) casing enclosing one or more parallel pulleys.

Block and tackle: system of two or more pulleys with a rope or cable threaded between them, usually used to lift or pull heavy loads.

Canopy anchor (CA): a method of anchoring a climbing or rigging line in the branches or canopy of a tree.

Cant hook: long-handled lever fixed with a blunt metal end to handle and roll logs; includes a swinging metal hook opposite the blunt end to create leverage.

Carabiner: aluminum or steel connecting device used in climbing and static rigging that is opened and closed by a spring-loaded gate.

Come-along: (1) portable winching device, using cable or rope to draw two objects closer together. (2) simple arrangement of rope knots and loops to create mechanical advantage.

Command-and-response system: system of vocal communication in tree care operations used to convey critical information and ensure understanding by another worker, often between a worker aloft and a ground worker.

Double-locking carabiner: one which requires two distinct motions to prepare the gate to open.

Drop/landing zone: predetermined area where cut branches or wood sections will be dropped or lowered from a tree, and where the potential exists for struck-by injuries.

Figure-8 descender: metal device used to provide friction on a rope in climbing or very light rigging.

Kickback: sudden, sometimes violent and uncontrolled backward or upward movement of a chain saw.

Load/lowering line: rope used to lower a tree branch or segment that has been cut.

Lowering device (friction device) : a piece of equipment that is attached to the tree base or other anchor in rigging. Used to take wraps with the load lines to provide friction for controlled lowering of sections of a tree.

Mechanical advantage (MA): system by which effort or force can be multiplied.

Personal protective equipment (PPE): personal safety gear such as helmet, safety glasses, hearing protection, gloves, and leg protection, including chaps.

Pole saw: long-handled tool with a pruning saw on the end, used to cut branches from a distance.

Port-a-Wrap: brand name of a device used to create and manage friction in the lowering line (load line) in rigging operations.

Pull line (tagline) : line attached near the top of a tree or tree section to be felled or any tree section being removed, to help pull or guide it in the desired direction or to control its swing.

Pulley: device consisting of a rotating, grooved wheel between two side plates or cheekplates. Used to change the direction of pull in a line, as a fair lead to guide a line, or as part of a mechanical advantage system (compare with *block*).

Prusik loop: loop of rope used to form a Prusik hitch for climbing or rigging.

Throwbag: weighted sack used to set climbing or rigging lines in trees. Usually a shot-filled, teardrop-shaped bag.

Throwline: thin, lightweight cord attached to a throwbag or throwing ball used to set climbing or rigging line in trees.

Rigging point: place in the tree (natural or false crotch) that the load line passes through to control limb removal in rigging operations.

Webbing sling (web sling) : length of sewn webbing, often formed into a loop, used as an attachment in rigging.

hoopie sling: sling with one fixed eye and one adjustable eye, made from hollow-
id rope.

k zone: defined area of a job site, marked with caution signs and/or cones,
potential hazards exist and safety measures are in place to avoid accidents.

of the terms were taken from the ISA Glossary of Arboricultural Terms, 2020.

End Notes

1. From an article in the January 2017 *Tree Care Industry Magazine* titled: "Changing Human Behavior Through Training, "by Tracy Hawks.
2. Adapted from the man's prayer on the *Red Green Show:* "I'm a man, I can change, if I have to, I guess." However, used in the context of substance abuse, it is no joking matter.
3. Donald F. Blair, *Arborist Equipment,* (International Society of Arboriculture, 1995), 22.
4. From C.S. Lewis's book, *The Lion, the Witch and the Wardrobe.* In Lewis's quote the "it" is Aslan the lion: "Who said anything about safe? 'Course he isn't safe. But he's good."
5. Some common "advanced industry credentials" include: ISA Certified Arborist and ISA Certified Tree Worker Climber Specialist (International Society of Arboriculture, ISA), and Certified Treecare Safety Professional (Tree Care Industry Association, TCIA).
6. ANSI Z89.1 is the safety standard that establishes the types and classes of protective helmets, depending on the type of hazard encountered.
7. The Golden Rule: "So whatever you wish that others would do to you, do also to them." Matthew 7:12, *The Holy Bible: English Standard Version.*
8. For an excellent downloadable guide for sharpening chain saws, go to www.stihl.com/sharpening-brochure.aspx
9. Brian J. Ruth, Jen W. Ruth, *Homeowner's Complete Guide to the Chainsaw,* (Fox Chapel Publishing Company, Inc., 2009), 187.
10. These driving habits were adapted from The National Safety Council's online Defense Driving Course. This would be a great course for all employees who drive work vehicles.
11. The SIR sequence was taken from the book *The Civility Solution,* by P.M. Forni, (St. Martin's Press, 2008), 71.
12. The movie *Master and Commander* was adapted in part from Patrick O'Brian's first historical nautical novel of the same title. O'Brian wrote over 20 books in this series. Arborist Tom Dunlap first introduced me to these books. He believed that O'Brian's books—as well as the movie—to be an excellent source for lessons on teamwork and leadership.
13. This almost happened at a climbing championship where I witnessed a competitor come within two feet of the end of his line during a descent. It was the yells and screams of the onlookers that got his attention and caused him to stop—still 30 feet above the ground.
14. The best book available on this topic is *The Art and Science of Practical Rigging,* by Peter Donzelli and Sharon Lilly, a publication of the International Society of Arboriculture.
15. Dr. Jerome J. Notkin, Sidney Gulkin, *The How and Why Wonder Book of Machines* (Grosset & Dunlap, 1960), 46.
16. Donald F. Blair, *Arborist Equipment,* (International Society of Arboriculture, 1995), 27
17. *Tree Care Industry Magazine* is published monthly by the Tree Care Industry Associatio
18. One such book is *To Fell a Tree,* by yours truly. See page 173 for others.
19. These three tricks were those of climbing guru Peter Jenkins: www.treeclimbing.co